Agricultural Extractivism in the Mediterranean Region

Susana Narotzky
Natalia Buier • Theodora Vetta
Editors

Agricultural Extractivism in the Mediterranean Region

A Socioecological View

palgrave
macmillan

Editors
Susana Narotzky
University of Barcelona
Barcelona, Spain

Natalia Buier
Complutense University of Madrid
Madrid, Spain

Theodora Vetta
University of Barcelona
Barcelona, Spain

ISBN 978-3-032-05598-9 ISBN 978-3-032-05599-6 (eBook)
https://doi.org/10.1007/978-3-032-05599-6

Cover illustration: Maram_shutterstock

This Palgrave Macmillan imprint is published by the registered company Springer Nature Switzerland AG.
The registered company address is: Gewerbestrasse 11, 6330 Cham, Switzerland

Acknowledgements

This volume was made possible by the collective research, debate and continuous exchanges that took place within the framework of the project "Valuable food, essential workers, vulnerable people, and social responses to crisis: food provisioning systems during the COVID-19 pandemic" PID2020-114317GB-I00 from the Spanish State Research Agency. While most of the group members are among the contributors to this collection, we would also like to express our gratitude to Alicia Reigada, Bibiana Martínez Álvarez and María Giménez Casalduero, whose inputs have been essential to our discussions and debates. The workshop "Europe and the Extraction-Exploitation Nexus: Studying Systemic Social Reproduction in Times of Environmental Crisis" (June 2023, Barcelona), organized by Natalia Buier and Doris Buu-Sao (Université de Lille), provided an opportunity for the editors as well as several of the contributors to deepen their reflections on the central topics of this volume. We would also like to express our gratitude to Jaume Franquesa, whose collaboration with the three of us, at different moments, has been a constant source of inspiration for thinking about the relationship between exploitation and extraction and so many other issues. We also wish to thank all the participants to the seminars of the Research Group on Reciprocity (GER) at the Universitat de Barcelona, who have constantly provided us with stimulating debates.

Competing Interests
Translation and publication of this edited collection has been supported by grant PID2020-114317GB-I00 from the Spanish State Research Agency.
Natalia Buier received support through the RYC2023-045872-I funded by MICIU/AEI/10.13039/501100011033 and by ESF+.
Theodora Vetta received support through the RYC2023-044674-I funded by MICIU/AEI/10.13039/501100011033 and by ESF+, and by the Independent Research Fund Denmark grant no. 2102-00037B.

Contents

Notes on Contributors

Gennaro Avallone holds a PhD in Sociology and Social Research and is Associate Professor in Sociology of Environment and Territory in the Department of Political and Social Studies at the University of Salerno. His main research areas are the following: international migrations, agricultural labour, socio-ecological relationships, and political ecology. He collaborates with the international research networks 'Agromig', 'World-ecology network', and 'Red de saberes compartidos'.

Irene Guerrero Avellaneda is a predoctoral researcher. She holds a degree in Sociology and a master's degree in Digital Society: Research and Analysis of Social Transformations from the University of Murcia (UMU). She is the recipient of the 2023 UMU Chair Award for Innovation and Participation and received a research collaboration grant for the 2023–2024 academic year. Her research interests include digitalization and agri-food systems.

Natalia Buier is a Ramón y Cajal researcher at the Complutense University of Madrid (UCM), a member of the Grup d'Estudis sobre Reciprocitat (University of Barcelona) and the Charles Babbage Social Sciences of Work research group (UCM). Her current research focuses on agriculture-conservation conflicts and the integration of social and environmental justice in a context of groundwater depletion.

Carmelo Buscema holds a PhD and is Senior Assistant Professor in Political Sociology at the University of Calabria, Italy. He carried out fieldwork and research activities in many countries and published several essays on international migrations, capitalism, cooperation and development, new ICTs, and geopolitics.

Davide Cacchioni is a political-economic anthropologist. He has carried out ethnographic research in northern Italy and southern France. His studies focus on agriculture, economic alternatives, (post)industrialism and political regulation, with a keen eye on the socio-ecological relationships in which socio-economic dynamics are forged. He holds a PhD from the EHESS (École des Hautes Études en Sciences Sociales - France).

Andrés Pedreño Cánovas is Professor of Sociology at the University of Murcia and editor of the journal *Sociología Histórica* published by the University of Murcia. His research focuses on rural sociology, labour, migrations and the environment. He has authored numerous books and specialised articles, most recently: *De Cadenas, Migrantes y Jornaleros: los territorios rurales en las cadenas globales agroalimentarias* (Madrid, 2014); and *Sociología rural y de la agricultura: nuevos avances en investigación social de los campos de la globalización agroalimentaria* (CIS, 2025).

Francesco Saverio Caruso is research fellow at the University of Calabria and a member of the Centre for the Study of Migration and Intercultural Relations at the University of Almería (Spain). His research interest revolves around migrant labour in agriculture, union and labour movements, and local policies for migrant inclusion.

Paula Escribano Castaño holds a PhD in Social Anthropology from the Autonomous University of Barcelona (UAB) and is assistant professor in the Department of Social Anthropology and Social Psychology at the Complutense University of Madrid (UCM), and a member of the Reciprocity Studies Group at the University of Barcelona (UB). Her main research interests lie in livelihood strategies within political ecology and political economy approaches, with a focus on agrarian systems and, more recently, on transhumant shepherds. She has directed and produced ethnographic films on rural transformation.

Juan Castillo-Rojas-Marcos holds a BSc in Sociology and Politics from the Universidad Pablo Olavide (UPO, 2020) and a master's degree in Social Anthropology from the University of Sevilla (US, 2023). He is pursuing a PhD in Migration Studies at the Instituto Universitario de Estudios sobre Migraciones (IUEM), Universidad Pontificia Comillas. He is technical editor of the journal *Migraciones* and secretary of the Research Committee on International Studies of the Spanish Federation of Sociology (FES). He studies the intersection between migration, class conflict, coloniality and the social production of space.

Alessandra Corrado is Associate Professor of the Sociology of Territory and Environment at the University of Calabria. She is a member of the Research Study Group 'Southern and Mediterranean Europe' in the European Society for Rural Sociology, and of RN38 'Space, Society and Rurality' Board in the European Sociological Association. Her research interest revolves around agri-food restructuring, international migrations, rural livelihoods, agrarian question and agroecology.

Dhouha Djerbi is a PhD Candidate at the Department of International Relations/Political Science at the Geneva Graduate Institute. Her dissertation research examines state-society relations in the Tunisian countryside, rural contentious politics, and gendered modes of claim-making in the post-uprising era.

Jordi Gascón holds a PhD in Social Anthropology and is an associate professor at the University of Barcelona. Previously he was attached to the University of Lleida and was senior researcher at the Instituto de Altos Estudios Nacionales (IAEN) in Ecuador. He specializes in rural studies and has conducted research in the Andean region and in the Iberian Peninsula. He is a member of the Food Action and Research Observatory (FARO), Research Institute of Nutrition and Food Safety (INSA-UB) and the research group Anthropology of Crises and Contemporary Transformations (CRITS).

Patricia Homs Ramírez de la Piscina is a postdoctoral researcher specializing in food systems and agroecology. She has been conducting research on alternative food networks in Catalonia for over ten years. Since 2013, she is a member of the Reciprocity Studies Group, and she is currently part of the Food Action and Research Observatory at the University of Barcelona.

Andrea Mathez her broad scholarly interest lies in the relationship between societies and their environments. She works in the fields of agri-food studies, critical development studies and political ecology. Her PhD research examines how (geo)politics and intimacy intersect in the context of alternative agricultures in Morocco and Switzerland. She also engages with questions of embodied epistemologies, situated solidarities and the ethics of international field research.

Antonio José Ramírez Melgarejo is Doctor in Sociology from the University of Murcia. Currently a researcher and professor at the Complutense University of Madrid, a member of the Charles Babbage Research Group in Social Labor Sciences and co-coordinates the Research Committee CI07 of the Spanish Federation of Sociology. He has published articles and book chapters in prestigious journals and participated in national and international seminars and conferences. His research focuses on the complex world of labour, strategies of social reproduction, labour migrations and the sociology of territory.

Yoan Molinero-Gerbeau is a political scientist with a PhD in Political Science and International Relations. He is a senior researcher at Instituto Universitario de Estudios sobre Migraciones (IUEM), Universidad Pontificia Comillas. He is editor-in-chief of the journal *Migraciones* and director of the Master's Degree in International Development Cooperation at Comillas University. His main lines of research are IR Structuralism, Critical Environmental Studies, and Migrant Worker's Conditions in Global Agriculture.

Susana Narotzky is Professor of Social Anthropology at the University of Barcelona, Spain. She received the Spanish National Award for Research in the Humanities in 2020. Her work seeks to produce a grounded model of social reproduction processes and is inspired by theories of critical political economy, moral economies, feminist economics and value regimes. She investigates valuation processes and valorization practices that affect social, economic and environmental sustainability, with a focus on food systems and energy transitions.

Michael Petrou is a social anthropologist and researcher at the National Centre for Social Research, Athens. His research interests and publications centre on an anthropological approach to rural space and rural-urban dynamics. He focuses particularly on the socio-economic transformation of the countryside and emerging challenges, such as social and environmental conflicts, land grabbing, multifunctionality and socio-spatial recomposition, consumption patterns, cultural identities, and the rise of fascist phenomena in rural areas.

Arianna Poletti is a PhD candidate in the national Italian programme on Climate Change and Sustainable Development (PhD-SDC) hosted by International School for Advanced Studies (IUSS Pavia). She has been based in Tunis for five years, working as an investigative journalist focusing on social and ecological issues in Tunisia. Her research applies a political ecology critique of the concept of climate mobility within the migration regime, examining the socio-political and environmental dynamics in the Mediterranean and North Africa.

Valeria Siniscalchi is Professor of Anthropology at the *École des Hautes Études en Sciences Sociales* (EHESS), member of the Research Center on circulations, links and exchanges (CeRCLEs, Marseille, France). Her research focuses on economic and political anthropology, and food activism. She has conducted extensive research in the North and South of Italy, in the French Alps and inside the Slow Food movement. Her recent books include *Food Values in Europe* (2019, with Krista Harper), and *Slow Food: The economy and politics of a global movement* (2023) published by Bloomsbury.

Cecilia Vergnano is a social anthropologist whose research interests include processes of borderization, racialization, dispossession and stigmatization—and related resistances—in the realms of mobility, labour and knowledge. She has been granted postdoctoral fellowships by the Research Foundation – Flanders (FWO) and *Fonds de la Recherche Scientifique* (FRS-FNRS) for her projects LABOR-PAS and AgriLaBor, respectively, on the intersection between border and labour regimes in the Italian agro-industrial sector.

Theodora Vetta is a Distinguished Researcher at the department of Social Anthropology, University of Barcelona, Spain. Her work focuses on green energy transition, financialization, labour and indebtedness in South Europe. She is co-editor at FOCAAL – *Journal of Global and Historical Anthropology*.

List of Figures

List of Tables

Beyond Depletion: On Value Extraction in Mediterranean Agriculture

Susana Narotzky, Natalia Buier, and Theodora Vetta

The present socioeconomic conjuncture—shaped by the aftermath of the COVID-19 pandemic, the war in Ukraine, and ongoing climate disruptions—has severely amplified pressures on both agricultural producers and households. Farmers face soaring input costs (for fuel, fertilizers, animal feed, and machinery), tightening environmental regulations, and unequal price competition, at a time when food affordability is increasingly critical for vulnerable working-class households grappling with inflation and stagnant or precarious incomes.

S. Narotzky (✉) • T. Vetta
University of Barcelona, Barcelona, Spain
e-mail: narotzky@ub.edu; theodora.vetta@ub.edu

N. Buier
Complutense University of Madrid, Madrid, Spain
e-mail: nbuier@ucm.es

S. Narotzky et al. (eds.), *Agricultural Extractivism in the Mediterranean Region*,
https://doi.org/10.1007/978-3-032-05599-6_1

Farmers' discontent has resulted in widespread demonstrations in many European countries and forced a reassessment of the "Green Deal" Common Agricultural Policies that aimed to regulate environmentally harmful practices and promote more sustainable ones, including the shift to agro-ecology, processes of land reclamation and rewilding. Simultaneously, the effects of climate change—heat, hydric stress—and of migratory policies create a particularly interesting relational context in the Northern and Southern shores of the Mediterranean where farmers, labor, and non-human nature articulate. The movement of agro-capital, people, and produce across borders, and the regulatory frameworks that design spaces of vulnerability, of scarcity, and of growth, are the background of the chapters in this collection and what creates a permanent conversation between them. The book connects spaces, people, and policies to agriculture, food provisioning, and economic prosperity.

Numerous studies have addressed food security in developing countries during famine situations caused by natural disasters, environmental crisis, economic downturns, or war (Baro & Deubel, 2006; Pingali et al., 2005; Food and Agriculture Organization, 2023; Mundy, 2018; Clapp & Moseley, 2020). Access to food has been addressed from the macro perspective of global food regimes where different practices of production, distribution, and consumption articulate across regions (McMichael, 2009; Friedmann, 2016), or from the micro perspective of proximity distribution and household consumption patterns mostly relating to class, gender, and age differential access, and to alternative agro-ecological cooperative practices (Caplan, 2016; Lambie-Mumford, 2013; Grasseni, 2013; Homs & Narotzky, 2019). Yet, few have addressed the connections between food provisioning, natural resource extraction, and the exploitation of human labor. In this edited collection, we wish to tackle the agricultural extractivist regimes in the Mediterranean, focusing on the socio-ecological and ideological practices that constrain or contribute to conflicting processes of social reproduction.

The volume is structured in two main sections, bringing together two faces of agricultural production around the Mediterranean. On the one hand, the extractive and exploitive relations that support an intensive export-oriented system of food provisioning. On the other hand, the possibilities and contradictions of sustainable and alternative practices of production and distribution that include digitalized and precision technologies, cooperative distribution, agro-ecological projects, short-circuit proximity provisioning, agro-ecology, as well as the expert frameworks that produce sustainable imagined futures. By addressing agriculture from this double perspective, we hope to provide elements for the analysis of complex realities that do not exist in separate realms but rather co-exist and are related at different scales.

The chapters address these issues from a multiscale perspective that approaches local practice and experience as inflected by institutional political economic structures, following Doreen Massey's injunction to analyze phenomena considering their "global sense of place" (1994). Selected cases from the Southern and Northern Mediterranean shores provide an integrated vision of movement and policies as they impact the production and availability of agricultural produce. Chapters are based on empirical research, with particular emphasis on ethnographic and qualitative methods as practiced in social anthropology, sociology, and human geography. The empirically grounded analysis provides theoretical insights into social reproduction crises that highlight the articulation of human and non-human valuations with economic valorization and capital accumulation.

Appropriation, Production and Reproduction: An Extractivist Dilemma

Why do we use extractivism to describe the kinds of relations that we find in a large part of Mediterranean agriculture? What is it that we seek to highlight? How does our perspective engage with other descriptions of agricultural extractivism and extractivism more generally?

A widely used definition of extractivism that goes beyond hydrocarbon and mining processes underlines the following characteristics: (1) extraction of primary commodities, that is, non-processed or minimally processed material produce; (2) high volume or intensity of extraction, environmental degradation; (3) over half of it for export; and (4) deterioration of labor opportunities or of labor conditions (Gudynas, 2015, McKay & Colque, 2021, for a recent overview, see Veltmeyer & Ezquerro-Cañete, 2023). These characteristics are conceived as part of a capitalist process whose main objective is to accumulate capital wealth. So why use extraction for a particular relation of capital with labor and nature instead of exploitation? Several arguments have been proposed for explaining the distinction.

The first—and theoretically central—refers to the process by which a benefit (or value) accrues to capital accumulation. In the case of industrial capitalism—the classic template for understanding value accumulation—production, through the process of combining different factors such as raw matter, technology, and labor, creates new value, often defined as "added" value. This process was understood in Marx's Labor Theory of Value as one of labor exploitation, where any additional value came from the differences between the creative power of embodied energy (labor), the market cost of labor power (the price of its reproduction) and other material inputs (means of production), and the socially necessary labor to produce commodities within an integrated exchange system (the Market). Many problems have been pointed out for this theory, from feminist and environmentally conscious Marxist scholars (Deléage, 1989; Dalla Costa & James, 1972; Federici, 2012), yet a solution to this problem has been to include as a central and ongoing form of capital accumulation what Marx appeared to reserve to an original phase of accumulation, "primitive accumulation." This is what Harvey (2003) has famously defined as accumulation by dispossession. This breakthrough underscored that in capitalist processes of accumulation an important and recurrent aspect is the appropriation of value beyond the production process; that is, the appropriation of value that is embodied and embedded in matter, people, or spaces that are not, or only partially, reproduced within the integrated exchange system of a capitalist market (Fraser, 2014, 2018). This form of value has been difficult to theorize within Marxist parameters and has given rise to many debates that we will not directly engage with (Burkett, 1996; Foster, 1999; Hornborg, 2020; Moore, 2011). Rather we wish to ponder on the unsettling question of value—what kind of value is appropriated in processes of dispossession—which is at the center of the concept of extractivism.

In a much quoted passage, Marx states that while nature has no value (embodied abstract socially necessary labor) since it has not been produced as a commodity within a capitalist process of production, it nonetheless has use-value and actively participates in the production of the use-value of commodities (Marx, 1976).[1] Does this mean that by entering

[1] "Use-values like coats, linen, etc., in short the physical bodies of commodities, are combinations of two elements, the material provided by nature, and labour. (...) Labour is therefore not the only source of material wealth, i.e. of the use-values it produces. As William Petty says, labour is the father of material wealth, the earth is its mother" (Marx, 1976, pp. 133–134). But "If an instrument of production has no value to lose, i.e. if it is not the product of human labour, it transfers no value to the product. It helps to create use-value without contributing to the formation of exchange-value. This is true of all those means of

a capitalist process of production, aspects of nature, matter, and energy become generative elements creating the material basis of value: use-value?[2] However, because they have not been fully "produced" in a capitalist process, their contribution to value as defined by "socially necessary labor" cannot be equated to the value emerging from production. Still, arguably, many inputs in production processes—such as energy, people, minerals, land, water, and knowledge—even as they enter the market as commodities, are not produced as such. Or, rather, they are not reproduced within capitalist relations of production, or only partially. Nevertheless, it would appear that they do contribute to value accumulation not only as free gifts of nature or as nature's basic use-value with which human labor interacts, whatever mode of production present. Here, Polanyi's concept of fictitious commodity gives us food for thought as it articulates a central contradiction: these means of production operate as commodities, yet they are not, meaning they are not produced as such (Polanyi, 1971). Rather, they are forced into becoming commodities through violent forms of expropriation. This has sometimes been theorized as a metabolical rift (Foster, 1999), where capitalist production and accumulation ruptures a previously balanced and integrated socio-ecological metabolism between humans and nature. In this reading, there is a historical "before and after" that separates human nature and capitalist processes from other-than-human nature and other-than-capitalist processes.

Yet, we wish to keep the ambivalence of the fictitious commodities and the continuous practices of violent expropriation that constitute them. In the process, there is a conversion and a capture that makes various kinds of resources and values commensurable within the circuit of capital. Violence is, then, the instrument for producing Polanyi's fiction—simultaneously a material constraint and an ideological hegemony. Colonialism and coloniality (Quijano, 2000) are emblematic of this logic, though it operates, more generally, in any capitalist process. From this perspective, extractivism is not an exceptional or peripheral feature of capitalism but an

production supplied by nature without human assistance, such as land, wind, water, metals in the form of ore, and timber in virgin forests" (Marx, 1976, p. 312).

[2] "With the exception of the extractive industries, such as mining, hunting, fishing (and agriculture, but only insofar as it starts by breaking up virgin soil), where the material for labour is provided directly by nature, all branches of industry deal with raw material, i.e. an object of labour which has already been filtered through labour, which is itself already a product of labour" (Marx, 1976, p. 287).

ever-present aspect of the concrete reproduction processes of production inputs or resources.

Ecological economists, such as Georgescu-Roegen (1975, see also Martínez Alier, 1993, Hornborg, 1998, Deléage, 1989) have delved into the problematic conversion of energy and matter into commodities. Georgescu-Roegen speaks of "economic myths" that push economists to "insist that resources are properly measured in economic, not physical terms", the "myth that the price mechanism can offset any shortages, whether of land, energy or materials" (Georgescu-Roegen, 1975, pp. 354–355). From the point of view of thermodynamics, "cost consists in essence of low entropy, not of money, and is subject to the limitations imposed by natural laws" (Georgescu-Roegen, 1975, p. 358). What appears relevant to our discussion of extractivism—as with the case of fictitious commodities—are the incommensurable aspects of physical and economic resources; specifically, how the accounting in terms of transforming low entropy—solar energy, earth matter, water—into high entropy—agricultural produce and waste—cannot be equated to market accounting mechanisms. In fact, the economic "myth" enables the appropriation of the energy-value in resources precisely because value is measured in terms of commodities, a market exchange criterion. The disparity expresses extraction: in terms of the low entropy–high entropy relation, intensive agriculture results in an overall loss of scarce low-entropy resources (environmental degradation), even as high productivity may increase overall market value. From this perspective as well, the concept of extractivism becomes applicable to all capitalist processes.

A crucial element in the definition of extractivism is tied to the export of resources, their removal from the regions of extraction to the regions where these resources will enter (industrial) production as inputs, forming the use-value basis for a process adding value through labor. Spatial separation participates in the conversion into commodities (commodification) of human and non-human resources and has been analyzed through the lens of unequal exchange and world-system theory's core-periphery relations. In his study on the Brazilian Amazon, Bunker (1984) explains the region's "relative underdevelopment" through successive waves of resource extraction (lumber, rubber, minerals) destined to industrial centers albeit not producing local (industrial) development. His thesis rests on the evidence that there are continuous unbalanced flows of matter and energy moving from "extractive peripheries" to "productive cores." As he puts it, "production requires the transformation of matter and energy which neither capital nor labor can create. Production cannot occur without extraction, and extractive systems function differently, and with different costs, than productive systems" (1984, p. 1018). In addition to human labor, Bunker (1984, p. 1028) describes these costs as including "human life, social organization and technology, and the ecological viability of various interdependent plant and animal systems" on which human communities depend. The crucial difference between extractive and productive economies, then, lies in the reproduction of resource inputs and, in the last instance, in the role of labor in the generation of value. Bunker is credited with proposing the concept of "modes of extraction"—a systemic connection of socio-ecological phenomena—as distinct but interdependent with "modes of production," highlighting that exchange inequalities express the fact that "value in nature is appropriated in one region and labor value incorporated in another" (1984, p. 1053). Movement is important and so is the form of value incorporated in commodities. The distinction between extraction and production follows the distinction between nature-value and labor-value and underscores the difference between distributive and generative forms of value accumulation (also present in the concepts of accumulation by dispossession and expanded accumulation). The concept of neo-extractivism has critically addressed the arguments of "pink tide" governments for presenting extractivism as a strategy for national development and forwarding redistributive agendas (Petras & Veltmeyer, 2014; Svampa, 2019). But while proponents of this model of accumulation have defended it as a way out of underdevelopment, production and extraction still appear as distinct modes of value accumulation.

Gudynas (2015) also follows this distinction by defining extractivisms as practices that are not "modes of production" but "modes of appropriation." This contrast has wider implications predominantly regarding the question of labor. Indeed, the theorization of extractivism often hinges on defining flows of value accruing to capital that are not created by labor in the production process but pertain to another domain of value. According to this view, extractive processes of any kind (including monocrop agriculture for export) transfer the nature-value of the primary commodities but require a low labor input. Hence, the argument goes, the term "industry" is a misnomer: what accrues value is not the labor-capital relation but the nature-capital one instead. Arguably, in order to export primary commodities (a key element of the definition), infrastructures have to be built requiring only temporary labor input. Likewise, although some extractive industries such as mining may initially require intensive labor, this is soon replaced by machinery and open strip mining. Many monocrop plantations also operate with proportionally few workers relative to the value extracted.

However, this take on extractivism appears particularly problematic. Many empirical cases contradict the assumption that value is generated through a proportionally low input of labor. Classical plantation systems, for instance, thrive on a large—albeit mostly unfree—labor force, and, in some cases—such as sugarcane—an industrial aspect is an inherent part of the local system of extraction (Mintz, 1986). Other examples such as small informal mining (Peluso, 2018) or contract farming (Little & Watts, 1994; Martiniello & Azambuja, 2019; Bahati et al., 2022) can be highly labor intensive while still conform to many of the defining elements of extractivism: minimally processed primary produce, intense extraction and environmental degradation, local social disruption, export-oriented production through cooperatives or other collection and commercialization systems, and poor labor conditions. Yet, what seems to characterize these extractive processes is not the absence of labor but its systematic devaluation: the cost of its reproduction tends to a minimum whether through slave and bonded labor, informal low wage labor, or family labor, and is often devalued socially and culturally. Hence, it is not so much that labor is irrelevant—justifying the non-"industrial" label of extractivism—but rather that the reproduction of labor, as that of nature, is not accounted for, particularly when it is informal, unpaid, or falls outside the wage labor condition (as with family work or help). Jason Moore's framework of capitalism's production of "cheap" natures provides a useful perspective here. He argues that resources are socially and historically produced through power-knowledge and exchange relations and in the process of the expansion of capitalism (commodity cash nexus)—in frontier spaces—, some forms of energy and matter are captured from outside the circuits of reproduction tied to the cash nexus. These resources—what Moore calls the "four cheaps" (labor, energy, food, and raw materials)—cannot be properly valued but are directly implicated in commodity production and exchange (Moore, 2015). What seems crucial here for our debate on extractivism is the historical and relational understanding of what becomes a "resource" for capitalist production, and the spatialized idea of "commodity frontiers."

Rasmussen and Lund (2018) have defined frontiers as "the discovery or invention of new resources," that is, "novel configurations of the relationship between natural resources and institutional orders that happen at particular moments in particular places" (2018, p. 388). In their view, frontier spaces are transitional spaces where the nature of resources is invented or reworked and they are in tension with territorialization strategies which aim at creating regimes of resource control (2018, p. 389). These processes generate new valuation systems about the environment, people, and knowledge that are concrete and localized while simultaneously connected to wider scales of political economy and socio-ecological spaces. The tension between the relative indeterminacy of resource frontiers and the purposeful aim to configure territorial control defining "access" (Ribot & Peluso, 2003) results in valuation struggles waged in terms of symbolic and physical violence. These frontier spaces of commodification have been described for extraction and conservation which both share the evacuation of people as a general premise: devalued people, revalued resources (Scott & Smith, 2017). Land grabs meant to produce biofuel plantations, and conservation zones meant to offset decarbonization through ecosystem services (e.g. carbon sequestration) which can be defined as forms of "green" extractivism, have expanded dispossession of natural and social communities on the grounds of an alleged global benefit (Feeney, 2024; Fairhead et al., 2012; Blake et al., 2023). This takes place through valuation struggles about the knowledge criteria defining what is of value and how to measure it that occur in resource frontiers. Robertson (2011, p. 397) has argued that it is "the struggle over the creation of value-bearing abstractions from its [nature's] materiality" that is the process that explains capitalist value capture, the violence of categorical abstractions supports circulation and appropriation through the production of fictitious commodities. As a result, it helps configure extractivist territories.

These general approaches to extractivism point to the tension between capital accumulation and (non)reproduction—in material, social, cultural terms—that scholars have grappled with for over a century. Central to it is the struggle about what constitutes valuable resources, who can access them, who will benefit from them, how and to what end. In the following section, we turn more specifically to agrarian extractivism.

Agrarian Extractivism

In what follows, we aim to connect some of these broader considerations on extractivism to the specific case of agrarian extractivism. The initial theorizing of agrarian extractivism occurred in conjunction to the analysis of neo-extractivism more generally, meaning that its defining characteristics were also understood to be connected to the volume of material extracted, the intensity of ecological impacts, and its export orientation (Gudynas, 2015). Debates around agrarian extractivism were triggered by the specific conjuncture of the early 2000s, one of accelerated land grabbing, soaring commodity prices, and deepening corporate control (Veltmeyer & Ezquerro-Cañete, 2023). Flex crops such as soy and oil palm became the paradigmatic cases for a type of agriculture considered to be, like other extractive industries, characterized by a tendency to deplete its own resource base (Petras & Veltmeyer, 2014; Ye et al., 2020).

More recent studies of agricultural extractivism have tried to overcome the limitations of this restricted view focused on the quantitative characteristics of a narrowly understood production process and analyze the social relations and power differentials constitutive of it (Chagnon et al., 2022; Veltmeyer & Ezquerro-Cañete, 2023). McKay et al. (2021) have proposed to formalize this expanded notion of extractivism by tracing its features along seven dimensions: sectoral particularities, capital flows, labor dynamics, resource access and property dynamics, knowledge flows, non-human energy and material flows, and, finally, territorial restructuring and its developmental effects. This move to an expanded notion of extractivism, one looking to tease out the power relations that constitute it, has opened the door to questions and lines of inquiry on which we draw at the same time as we try to advance them.

This incipient, but growing corpus of studies that draw on an expanded notion of agrarian extractivism is characterized, in our reading, by a stress on the expropriation of nature, questions of social reproduction, and depletive dynamics. While analyses have progressively incorporated a focus on labor, they have done so in a particular manner, we argue. If initial theorizations focused almost exclusively on the appropriation of non-human nature, more recently it has been argued that the deterioration of labor opportunities and labor conditions is another feature of it (McKay, 2017; McKay & Colque, 2021). While this modifies the conceptual scope, such understanding of the relationship between agrarian extractivism and labor still highlights its character as a distributive mode of value accumulation, rather than a generative one. Building on this premise, analysts have disputed the industrial character of agricultural extractivism, arguing that it does not actually lead to the industrialization of the countryside since it fails to generate employment, value-added processing, and sectoral linkages (McKay, 2017). Following an implicit—yet normative—understanding of industrialization as a motor of development, the above assumptions on the non- or anti-industrial character of agricultural extractivism fit within its broader diagnosis as a "counter-reproductive economy" (Alonso-Fradejas, 2021a), one that undermines its very conditions of continuity: an economy that impedes the alleged, positively valued, industrial development of productive forces, is driven by systematic spoliation (Andrade, 2022), and organizes life through "subjugation, depletion, and non-reciprocity" (Chagnon et al., 2022).

This focus on spoliation, predation, and depletion also seeps into the way questions of social reproduction have been addressed. On the one hand, some authors have explicitly characterized agrarian extractivism as "production without reproduction," a form of "resource-mining" (Ye et al., 2020) that inexorably leads to resource exhaustion, and as a "predatory, life-purging" form of accumulation (Alonso-Fradejas, 2021b). On the other hand, a different approach has been coalescing in those studies that address the destructive relations between economic production and social reproduction, where the latter is mostly understood as the sum of practices through which human and non-human lives are reproduced, including domestic work, subsistence agriculture, as well as nature's regenerative capacities (Mintz, 1986). In this direction, the acknowledged need to focus on the entanglements of production and reproduction (León Araya & Montoya, 2025), reveals the "extra-economic means through which the extractive capture of value is attained" and the central role that gender dynamics play in this (Ojeda, 2021, p. 87). When juxtaposing debates on the non-industrial character of agricultural extractivism with those concerning the relationship between production and reproduction, León Araya's diagnosis is confirmed: extractivist dynamics have predominantly been analyzed through a focus on destruction—of both production forces and conditions of reproduction (León Araya, 2021).

This volume shares with the above-mentioned approaches a concern with both the status of labor and social reproduction. However, we wish to bring a specific perspective to both these aspects. Regarding the relationship between labor and agrarian extractivism, we depart from approaches that define agrarian extractivism primarily as a model of accumulation characterized by the destruction of employment. While we acknowledge the centrality of diverse forms of labor devaluation to its operations, contemporary agricultural extractivism is defined, as the contributions in this volume attest, not only by the destruction of specific labor relations but also by the continuous formation of new laboring populations. The specific organization of the labor process may sometimes obscure the central role of labor, both waged and unwaged, within agricultural production. Yet, a grounded ethnographic perspective reveals that the contemporary reproduction of agrarian capital is inexorably dependent on the channeling of human energy into specific forms of exploitable labor.

Regarding the relationship between production and reproduction, we fully embrace those perspectives that insist on expanding the analysis to include questions of social reproduction. We do, however, insist that the particular way in which reproduction is brought to bear on production matters. While we acknowledge the importance of contributions that explore the entanglement of production and reproduction, highlight the role of non-market relations in reproducing capitalist structures of accumulation, and, more generally, bring the critique of the production-reproduction dichotomy into the study of agrarian extractivism, our perspective goes further by adopting social reproduction as an integral analytical framework. This means that, while we are sympathetic to attempts to overcome the production-reproduction dichotomy, we contend that perspectives that begin by treating production and reproduction as distinct realms, or that theorize reproduction through the "extra-economic," are ultimately insufficient to account for the ways in which agricultural extractivism channels life into the search for profit. Building on long-term work on the anthropological study of value and valuation (Narotzky, 1997, 2022), we maintain that reproductive processes should be understood as an integral part of economic processes, and vice versa. This requires two distinct moves: on the one hand, an analytical disaggregation of reproduction into distinct forms, and, on the other hand, a grounded analysis of their contradictory but inescapable unity (Narotzky, 1997). Agricultural extractivism offers a privileged window onto the contradictory unity of distinct reproductions, foregrounding the ways in which practices and relations through which capital and labor are reproduced are always embedded into processes of ecological reproduction.

As a result, we place value at the center of agrarian extractivism, and treat it as a process of channeling both human and non-human nature into pathways to accumulation. Agricultural extractivism, as a historically concrete form of producing capitalist landscapes, is, as we have argued elsewhere, "a key means of organizing and making mutually consistent the social division of nature and the social division of labor" (Buier & Franquesa, 2022, p. 12). Through it, the unity of nature-value and labor-value is reaffirmed.

Delving into the historical geographies of contemporary Mediterranean agriculture allows us to substantiate León Araya's observation according to which extractivist dynamics, predominantly analyzed as destructive, are actually also productive, and they suppose the incessant production of unstable more-than-human-orderings (León Araya, 2021). Our reading of the relevant literature and the analysis of the cases brought together in this volume both drive us to agree with the fact that the productive aspect of agrarian extractivism has been insufficiently studied and theorized. We understand the productive aspect to mean that wherever capital visits destruction upon territories, it also rebuilds spaces and places in the service of accumulation. It is only from within the actually existing geographies of agricultural extractivism that we can fully understand what this mode of capturing value entails. If extractivism "turns resources (be they natural or social) into finite products" (Ye et al., 2020), much of what is relevant about its agrarian form seems to us to occur precisely in the process of confronting depletion.

The latter accelerates capital's search for new deposits, be they of labor, water, or land, and this search is an active moment in transforming the geographies of accumulation. This entails, as it has long been argued, movement through space but also, as the chapters addressing migrant livelihoods attest, spatial strategies that work through a diversity of bordering practices. But it also supposes displacements that are not immediately recognizable as spatial in character, such as the symbolic devaluation of feminized laborers or livelihood practices that are seen as inimical to accumulation. Finally, we notice that the productive aspect of agricultural extractivism, the process of bringing together into specific configurations the accumulation of nature-value and labor-value, requires the constant intervention of the state as "the chief coordinator of exchangeability" (Buier & Franquesa, 2022, p. 13). Whether we are talking about land, people, or knowledge, the process of making them into commodities requires the constant intervention of the state for keeping in place those frameworks, legal, economic, institutional, through which exchangeability is obtained and ultimately capital accumulation can proceed.

Agrarian Extractivism in the Mediterranean

Agrarian extractivism in the Mediterranean is understood here as the result of historical processes of *longue durée* that have shaped the region's integration into global agro-food markets. Since the colonial and post-imperial restructurings of the nineteenth and early twentieth centuries, Mediterranean agriculture has progressively turned to export-oriented food production for serving external markets. Export crops such as cotton, raisins, tobacco, citrus, and olives were initially produced under colonial rule or foreign investment, and, more recently, the intensive production of crops such as fruits and vegetables mostly destined to European consumption, has been facilitated by EU's Common Agricultural Policy (CAP) and trade agreements with North African countries (Hamouchene, 2022; Riachi & Matiniello, 2023). As is always the case, such transformative processes are neither homogeneous nor linear but always expressions of tensions and contradictions, following multiscalar entanglements of path dependencies, political agency, accumulation strategies, and their respective temporalities. Despite several "reparative" efforts, as this volume demonstrates, current Mediterranean extractivism lies on a historically rooted socio-ecological regime of exploitation, dispossession, and exhaustion of human and non-human natures, shaped by both colonial legacies and more recent neoliberal restructuring.

This book is divided in two main parts: (1) labor, nature, and capital in food production: practices of extraction and exploitation; and (2) agricultural projects for a sustainable future: visions and contradictions, investigating the two faces of agricultural production and opening the field of their inter-connections.

The first three chapters take the reader to the shores of North Africa and the realm of appropriation of other-than-human natures, namely, water and land. Focusing on extractivist policies leading to resource depletion and their unequal distribution, they explore the consequences of export-agriculture not only in its ecological dimensions but also in the very possibilities of making a living in its margins. Past history is ubiquitous in these chapters that analyze present-day agricultural extractivism as a development of previous colonial economic and political relations. They underline the historical aspect of contemporary forms of exploitation, appropriation, and unequal exchange supporting extractive relations everywhere.

Andrea Mathez offers a grounded analysis of Morocco's Green Morocco Plan and the Strategy Green Generation, promising to transform the environmental and climatic crisis into an opportunity for economic growth and rural poverty reduction. On the one hand, the chapter demonstrates how such policies have intensified the already existing extractivist model of agriculture, further amplifying localized irrigation infrastructure, and subsidizing oasis extensions for new palm plantations. On the other hand, the chapter looks into alternatives to green modernization projects, struggling to survive while subjected to various pressures: from water scarcity, to real estate markets and "irrational" consumption demands of growing agroecological clients. The next chapter goes beyond discourses of water scarcity and state failure, exploring the fragmented—yet selective—water governance practices in Tunisia, what the author calls "uneven hydraulic stateness." Dhouha Djerbi masterfully traces the historical and structural transformations of water management from the colonial era, to the currently promoted export-led agricultural industry, before turning to the often gendered everyday water struggles of households, characterized by hydric unpredictability and the penalization of previously tolerated subsistence strategies. The next chapter, with a shared regional focus, tackles related issues of land dispossession. Arianna Polletti and Gennaro Avallone's study of the export-oriented Deglet Nour date monocultures in Southern Tunisia unpacks the underlying transformations of land tenure regimes, from the pre-colonial land administration as commons, to its gradual privatization under French colonial rule and IMF's structural adjustment programs. Establishing a "commodity frontier," for serving mostly EU markets, such monocultures not only damage fragile oasis ecosystems but also prompt population displacements and turn former collective farmers into cheap wage labor, often over-represented by precarious, left-behind female agricultural workers.

Susana Narotzky and Natalia Buier's chapter theoretically explores the relation between predation and accumulation, by looking at the production of labor and water as agricultural inputs in the strawberry enclaves of Huelva, Spain. The joint analysis of the extreme exploitation of migrant labor and water depletion based on over-irrigation and politically induced speculative water claims by farmers builds an expanded frame of agricultural extractivism, where destruction is predicated on the material and ideological expelling of reproduction from the valorization process. The following chapter delves deeper into the mechanisms of labor intermediation in the agricultural sector of southern Italy. Cecilia Vergnano's ethnography grounds the coexistence of different—yet complementary—political rationalities, approaching migrant labor as a favorable productive input, as a security problem, or as a victim of abuse demanding protective, humanitarian intervention. By looking at arrival infrastructure around housing provision, she shows the contradictory logics and practices of the criminalized *caporalato* system and the more recent emergence of non-profit organizations and other forms of intermediation by resident migrants, both straddling between protection and value extraction.

The final chapter in this section zooms out of the ethnographic analytical level in order to offer a broader view of the spatial dimensions of agro-industry's mode of appropriation. Juan Castillo-Rojas-Marcos, Yoan Molinero-Gerbeau, and Gennaro Avallone sketch a comprehensive frame for understanding agricultural extractivism in Southern Europe as a producer of social space. Looking at several interconnected layers—from ecosystems, to built environments, and symbolic spatial divisions—they show how the territorial reconfigurations and exclusionary foundations of agro-enclaves devour habitats.

The second part of the book opens with two chapters on digitalization efforts for offsetting the climate-environmental crisis. In their critical assessment of the pig farming industry in Murcia, Spain, Irene Guerrero, Andrés Pedreño, and Antonio J. Ramírez challenge the techno-optimistic corporate discourse that presents—and legitimizes—digitalization as a control mechanism of health and environmental risks. Instead, they demonstrate how digital tools that organize logistics, define and implement production times, or measure the spread and pace of waste disposal serve in practice as a form of technopolitics, reinforcing capital concentration and the logic of extraction of nature and labor, while reducing at the same time potential societal responses. The following chapter links similar digitalization trends to green extractivist processes in the "twin transition" of the Italian agrifood enclave of Foggia. Carmelo Buscema, Francesco Caruso, and Alessandra Corrado examine how quality certifications, automated harvesting, and digital platforms—for matching labor supply and demand, saving irrigation water, reducing pesticides—function as top-down techno-fixes that exacerbate the divide between small and large companies. Although designed to ensure sustainable practices and "ethical compliance" throughout the supply chain, digitalization also produces superfluous human labor and unqualified produce as waste.

The next three chapters address the challenges faced by several small-scale initiatives aimed at strengthening local food supply chains and agro-ecological projects. In a context of declining CAP support and market deregulation, Michael Petrou's ethnographic insights offer a clear picture of how petty-capitalist family farming navigates recurrent and multidimensional crises—such as the 2009 sovereign debt crisis or the current energy crisis—in Beotia, Greece. Despite the overexploitation of migrant labor, the so-called resilience of family farms seems increasingly shaken by spiking taxation, soaring input costs, and restricted credit access, rendering other "green investments," such as solar energy parks, more attractive. In Catalonia, Spain, agro-ecological food provisioning systems similarly struggle to maintain their sustainability. Patricia Homs and Paula Escribano analyze the production of vulnerability among small-scale farmers, food processors, and herders, with particular emphasis on their political positionings. Despite reviewing—though not without tensions—their previous rejection of relations with the state by radically resignifying subsidies as entitlements, extensive herding and agricultural systems remain suffocated by disrupted land access, bureaucratization, certified product standardization, and unpaid work. Finally, Davide Cacchioni's chapter examines the development of alternative "hemp supply chains" in the Susa Valley, Italy, as manifestations of ethical entrepreneurialism. In a region with a long history of mobilizations against expropriative mega-projects—such as the TAV high-speed railway, or alpine ski resorts—and despite financial and technological constraints, various economically marginal but ideologically cohesive hemp "niches" continue to inspire imaginaries of alternative, collaborative economies.

Last but not least, the book concludes with two chapters tackling producers' associational and cooperative practices around commercialization. Jordi Gascón's critical study of fruit cooperatives in Catalonia, Spain, shows how their initial goal to optimize production and strengthen producers' collective bargaining has been stalled by the logistical imperatives of large-scale distribution. Large retailers have in fact further consolidated their control over both production and commercialization by turning cooperatives into their supervisory agents, while outsourcing the very costs of quality controls. In response, several Catalan producers are turning to agro-ecology as a way to access more autonomous distribution channels. A comparable development is explored in the final chapter through the case of Associations for the Preservation of Peasant Agriculture (AMAP) in France. Valeria Siniscalchi analyzes the organizational and political logics behind these "solidarity contracts" between small-scale agro-ecological producers and risk-sharing, active consumers, tracing their expansion during the COVID-19 pandemic and their partial decline in its aftermath. Although the surge in interest for direct provisioning did not endure, temporary urban consumers were not left unaltered: such experiments often gave rise to new spaces and more sustainable ways of thinking about and engaging with food.

Overall, this collection brings together empirical cases from different shores of the Mediterranean, each shaped by different histories of agricultural intensification. However, despite the differences, they all present trajectories of resource creation, appropriation, and depletion, and the transfer of value to spaces of capital accumulation. In some cases—such as Tunisia and Morocco—the classical export-orientation model still persists: primary resources flow from periphery or dependent nations to economic centers or former metropolises. In other cases—such as Italy and Spain—the extractive movement is not nationally defined and outbound, as domestic value accumulation by private firms is also important. Yet, elements such as environmental degradation and deterioration of labor conditions appear prevalent across all cases. The chapters in the first part of the volume refer clearly to the appropriation of local resources and the transfer of value to centers of capital accumulation. In contrast, the chapters in the second part manifest the multiple and contradictory faces of "sustainability" projects, some supported by institutional frameworks geared to "greening" agriculture, others backed by local grassroots movements seeking an integral transformation of food provisioning systems through agro-ecology, cooperatives and short-circuit distribution. Nevertheless, even in these contexts, we notice elements of extractivism: in some cases very explicitly—such as through digitalization trends—and in others more subtly—as seen within farmers' commercialization cooperatives. With the exception of local agro-ecology projects, soil and water depletion, contamination, and degraded labor conditions—including self-exploitation—are persistent and widespread. In all these cases, moreover, political institutions play a crucial role in creating the limits and categories—such as certification—that enclose "legal sustainability" and condition agricultural production.

Two key elements of agrarian extractivism emerge from the cases in this collection. The first refers to the "commodity frontier" aspect of extractivism: a dynamic process through which resources are continuously created and depleted as new resources—or new uses for existing resources—are produced anew, in a struggle for defining value. This includes material and immaterial resources, human and non-human nature, energy, raw matter, primary commodities, and knowledge. This process encompasses private and public actors, bottom-up and top-down mobilizations, that seek to categorize, define, and access valued or devalued resources. The second element concerns the capacity of concrete human and non-human communities, and their socio-ecological connections, to reproduce themselves, that is, to sustain life over time. When the capture of value that serves capital accumulation transforms relations between people and the environment in ways that endanger, forcibly displace, disperse, or outright destroy living communities, we are faced with an extractivist process. In agrarian extractivism, these two processes are deeply intertwined and in a dialectic tension where the metabolic relation between humans and non-human nature appears mediated by the aims of capital.

References

Alonso-Fradejas, A. (2021a). 'Leaving no one unscathed'in sustainability transitions: The life purging agro-extractivism of corporate renewables. *Journal of Rural Studies, 81*, 127–138. https://doi.org/10.1016/j.jrurstud.2020.10.001

Alonso-Fradejas, A. (2021b). Life-purging agrarian extractivism in Guatemala: Towards a renewable but unlivable future? In B. M. Mckay, A. Alonso-Fradejas, & A. Ezquerro-Cañete (Eds.), *Agrarian extractivism in Latin America* (pp. 139–164). Routledge.

Andrade, D. (2022). Neoliberal extractivism: Brazil in the twenty-first century. *The Journal of Peasant Studies, 49*(4), 793–816. https://doi.org/10.1080/03066150.2022.2030314

Bahati, I., Martiniello, G., & Abebe, G. K. (2022). The implications of sugarcane contract farming on land rights, labor, and food security in the Bunyoro subregion, Uganda. *Land Use Policy, 122*, 1–13. https://doi.org/10.1016/j.landusepol.2022.106326

Baro, M., & Deubel, T. F. (2006). Persistent hunger: Perspectives on vulnerability, famine, and food security in Sub-Saharan Africa. *Annual Review of Anthropology, 35*, 521–538. https://doi.org/10.1146/annurev.anthro.35.081705.123224

Blake, L. J., Chohan, J. K., & Escobar, M. P. (2023). Agro-extractivism and neoliberal conservation: *campesino* abandonment in the Boyacá páramos, Colombia. *Journal of Rural Studies, 102*, 103071. https://doi.org/10.1016/j.jrurstud.2023.103071

Buier, N., & Franquesa, J. (2022). Producing capitalist landscapes: Ethnographies of the green transition and its contradictions. *Capitalism Nature Socialism, 33*(4), 5–17. https://doi.org/10.1080/10455752.2022.2138481

Bunker, S. G. (1984). Modes of extraction, unequal exchange, and the progressive underdevelopment of an extreme periphery: The Brazilian Amazon, 1600–1980. *American Journal of Sociology, 89*(5), 1017–1064.

Burkett, P. (1996). Value, capital and nature: Some ecological implications of Marx's critique of political economy. *Science & Society, 60*(3), 332–359. https://www.jstor.org/stable/40403575

Caplan, P. (2016). Big society or broken society? Food banks in the UK. *Anthropology Today, 32*(1), 5–9. https://doi.org/10.1111/1467-8322.12223

Chagnon, C. W., Durante, F., Gills, B. K., Hagolani-Albov, S. E., Hokkanen, S., Kangasluoma, S. M., & Vuola, M. P. (2022). From extractivism to global extractivism: The evolution of an organizing concept. *The Journal of Peasant Studies, 49*(4), 760–792. https://doi.org/10.1080/03066150.2022.2069015

Clapp, J., & Moseley, W. G. (2020). This food crisis is different: COVID-19 and the fragility of the neoliberal food security order. *The Journal of Peasant Studies*, Published online: October 11 2020. https://doi.org/10.1080/03066150.2020.1823838.

Dalla Costa, M., & James, S. (1972). *The power of women and the subversion of the community.* Falling Wall Press; Butler and Tanner, Ltd.

Deléage, J.-P. (1989). Eco-marxist critique of political economy. *Capitalism Nature Socialism, 1*(3), 15–31. https://doi.org/10.1080/10455758909358382

Fairhead, J., Leach, M., & Scoones, I. (2012). Green grabbing: A new appropriation of nature? *Journal of Peasant Studies, 39*(2), 237–261. https://doi.org/10.1080/03066150.2012.671770

Federici, S. (2012). The reproduction of labor power in the global economy and the unfinished feminist revolution; women, land struggles, and globalization: An international perspective. In S. Federici, *Revolution at point zero: Housework, reproduction, and feminist struggle*, (pp. 91–115 and 126–138), PM Press.

Feeney, J. K. (2024). Biodiversity conservation under green extractivism and armed neoliberalism in Colombia. *Journal of Political Ecology, 31*, 464–487. https://doi.org/10.2458/jpe.5495

Food and Agriculture Organization. (2023). *The impact of disasters on agriculture and food security 2023—Avoiding and reducing losses through investment in resilience.* Food and Agriculture Organization United Nations. https://doi.org/10.4060/cc7900en

Foster, J. B. (1999). Marx's theory of metabolic rift: Classical foundations for environmental sociology. *American Journal of Sociology, 105*(2), 366–405. https://doi.org/10.1086/210315

Fraser, N. (2014). Behind Marx's hidden abode. *New Left Review, 86*, 55–72.

Fraser, N. (2018). Roepke lecture in economic geography—From exploitation to expropriation: Historic geographies of racialized capitalism. *Economic Geography, 94*(1), 1–17. https://doi.org/10.1080/00130095.2017.1398045

Friedmann, H. (2016). Commentary: Food regime analysis and agrarian questions: Widening the conversation. *The Journal of Peasant Studies, 43*(3), 671–692. https://doi.org/10.1080/03066150.2016.1146254

Georgescu-Roegen, N. (1975). Energy and economic myths. *Southern Economic Journal, 41*(3), 347–381. https://doi.org/10.2307/1056148

Grasseni, C. (2013). *Beyond alternative food networks: Italy's solidarity purchase groups*. Bloomsbury.

Gudynas, E. (2015). *Extractivismos. Ecología, economía y política de un modo de entender el desarrollo y la Naturaleza*. Centro de Documentación e Información Bolivia (CEDIB).

Hamouchene, H. (2022). The political ecology of extractivism in North Africa. In H. Veltmeyer & P. Bowles (Eds.), *The essential guide to critical development studies* (pp. 327–334). Routledge.

Harvey, D. (2003). *The new imperialism*. Oxford University Press.

Homs, P., & Narotzky, S. (2019). Within and beyond the market system. A case of organic food cooperatives in Catalonia. In K. Harper & V. Siniscalchi (Eds.), *Food values in Europe: Economies, ideologies, and power in practice* (pp. 132–146). Bloomsbury.

Hornborg, A. (1998). Towards an ecological theory of unequal exchange: Articulating world system theory and ecological economics. *Ecological Economics, 25*, 127–136. https://doi.org/10.1016/S0921-8009(97)00100-6

Hornborg, A. (2020). Towards an ecological theory of unequal exchange: Articulating world system theory and ecological economics. *Journal of World-Systems Research, 26*(2), 184–202. https://doi.org/10.1016/S0921-8009(97)00100-6

Lambie-Mumford, H. (2013). 'Every town should have one': Emergency food banking in the UK. *Journal of Social Policy, 42*(1), 73–89. https://doi.org/10.1017/S004727941200075X

León Araya, A. (2021). Agrarian extractivism and sustainable development: The politics of pineapple expansion in Costa Rica. In B. M. Mckay, A. Alonso-Fradejas, & A. Ezquerro-Cañete (Eds.), *Agrarian extractivism in Latin America* (pp. 99–116). Routledge.

León Araya, A., & Montoya, V. (2025). The border pineapple plantation economy: Labor and social reproduction in the North of Costa Rica. *Agriculture and Human Values*, 1–11. https://doi.org/10.1007/s10460-025-10714-3

Little, P. D., & Watts, M. J. (1994). *Living under contract: Contract farming and Agrarian Transformation in Sub-Saharan Africa*. University of Wisconsin Press. https://doi.org/10.2458/v6i1.21519

Martínez Alier, J. (1993). Valoración económica y valoración ecológica. In J. M. Naredo & F. Parra (Eds.), *Hacia una ciencia de los recursos naturales* (pp. 29–56). Siglo XXI.

Martiniello, G., & Azambuja, R. (2019). Contracting sugarcane farming in global agricultural value chains in Eastern Africa: Debates, dynamics, and struggles. *Agrarian South: Journal of Political Economy, 8*(1–2), 208–231. https://doi.org/10.1177/2277976019851955

Marx, K. (1976 [1867]). *Capital. A critique of political economy* (Vol. 1. B. Fowkes, Trans.). Penguin Books / New Left Review.

Massey, D. (1994). A global sense of place. In D. Massey (Ed.), *Space, place and gender* (pp. 146–156). Polity.

McKay, B. M. (2017). Agrarian extractivism in Bolivia. *World Development, 97*, 199–211. https://doi.org/10.1016/j.worlddev.2017.04.007

McKay, B. M., Alonso-Fradejas, A., & Ezquerro-Cañete, A. (2021). Introduction. In B. M. Mckay, A. Alonso-Fradejas, & A. Ezquerro-Cañete (Eds.), *Agrarian extractivism in Latin America* (pp. 1–20). Routledge.

McKay, B. M., & Colque, G. (2021). Extractive dynamics of agrarian change in Bolivia. In B. M. McKay, A. Alonso-Fradejas, & A. Ezquerro-Cañete (Eds.), *Agrarian extractivism in Latin America* (pp. 45–63). Routledge.

McMichael, P. (2009). A food regime analysis of the 'world food crisis'. *Agriculture and Human Values, 26*, 281–295. https://doi.org/10.1007/s10460-009-9218-5

Mintz, S. W. (1986). *Sweetness and power. The place of sugar in modern history.* Penguin. https://doi.org/10.1017/S0395264900069456

Moore, J. W. (2011). Transcending the metabolic rift: A theory of crises in the capitalist world-ecology. *The Journal of Peasant Studies, 38*(1), 1–46. https://doi.org/10.1080/03066150.2010.538579

Moore, J. W. (2015). Capitalism in the web of life. In *Ecology and the accumulation of capital.* Verso. https://doi.org/10.2458/v24i1.21075

Mundy, M. (2018). *The strategies of the coalition in the Yemen war: Aerial bombardment and food war.* World Peace Foundation.

Narotzky, S. (1997). New Directions in Economic Anthropology, London: Pluto Press.

Narotzky, S. (2022). To have a life: Labor reproduction, value and negative value. In S. Kasmir & L. Gill (Eds.), *The Routledge Handbook of the Anthropology of Labor (pp.3-16). London: Routledg*

Ojeda, D. (2021). Social reproduction, dispossession, and the gendered workings of agrarian extractivism in Colombia. In B.M, Mckay, A. Alonso-Fradejas, & A. Ezquerro-Cañete (Eds.), *Agrarian Extractivism in Latin America (pp. 86-98). London: Routledge.*

Peluso, N. L. (2018). Entangled territories in small-scale gold mining frontiers: Labor practices, property, and secrets in Indonesian gold country. *World Development, 101*, 400–416. https://doi.org/10.1016/j.worlddev.2016.11.003

Petras, J., & Veltmeyer, H. (2014). *Extractive imperialism in the Americas. Capitalism's new frontier*. Brill.

Pingali, P., Alinovi, L., & Sutton, J. (2005). Food security in complex emergencies: Enhancing food system resilience. *Disasters, 29*(S1), S5–S24. https://doi.org/10.1111/j.0361-3666.2005.00282.x

Polanyi, K. (1971 [1944]). *The great transformation*. Beacon Press.

Quijano, A. (2000). Coloniality of power and Eurocentrism in Latin America. *International Sociology, 15*(2), 215–232. https://doi.org/10.1177/0268580900015002005

Rasmussen, M. B., & Lund, C. (2018). Reconfiguring frontier spaces: The territorialization of resource control. *World Development, 101*, 388–399. https://doi.org/10.1016/j.worlddev.2017.01.018

Riachi, R., & Matiniello, G. (2023). Manufactured regional crises: The Middle East and North Africa under global food regimes. *Journal of Agrarian Change, 23*(4), 792–810. https://doi.org/10.1111/joac.12547

Ribot, J. C., & Peluso, N. L. (2003). A theory of access. *Rural Sociology, 68*(2), 153–181. https://doi.org/10.1111/j.1549-0831.2003.tb00133.x

Robertson, M. (2011). Measurement and alienation: Making a world of ecosystem services. *Transactions of the Institute of British Geographers*, NS, *37*, 386–401. https://doi.org/10.1111/j.1475-5661.2011.00476.x.

Scott, D., & Smith, A. A. (2017). Sacrifice zones in the green energy economy: The new climate refugees. *Transnational Law and Contemporary Problems, 26*(2), 371–382. https://doi.org/10.7202/1042776ar

Svampa, M. (2019). *Neo-extractivism in Latin America: Socio-environmental conflicts, the territorial turn, and new political narratives*. Cambridge University Press. https://doi.org/10.1017/9781108752589

Veltmeyer, H., & Ezquerro-Cañete, A. (2023). Agro-extractivism. *The Journal of Peasant Studies, 50*(5), 1673–1686. https://doi.org/10.1080/03066150.2023.2218802

Ye, J., Van Der Ploeg, J. D., Schneider, S., & Shanin, T. (2020). The incursions of extractivism: Moving from dispersed places to global capitalism. *The Journal of Peasant Studies, 47*(1), 155–183. https://doi.org/10.1080/03066150.2018.1559834

LABOR, NATURE AND CAPITAL IN FOOD PRODUCTION: PRACTICES OF EXTRACTION AND EXPLOITATION

Thirsty Landscapes: Morocco's Green Agricultural Policies and the Struggle for Water at its Margins

Andrea Mathez

Introduction

World food prices for staples like rice, wheat, and maize spiked in 2007, triggering the 2007–2008 World Food Crisis (WFC), which severely affected many countries in the 'global South'. Although food security is a multi-dimensional concept, of which food availability is just one component, the WFC has led to a renewed framing of food insecurity as a technical problem that can be resolved through market regulations and increased productivity. This has encouraged new efforts by scientists, states, and donors to start a 'New Green Revolution', as reflected in the birth of the Alliance for a Green Revolution in Africa (AGRA) in 2008. Among the many initiatives unfolding under AGRA, the Moroccan government launched in 2007 the Green Morocco Plan (GMP), followed in 2020, by

A. Mathez (✉)
Institute of Geography and Sustainability, University of Lausanne, Lausanne, Switzerland
e-mail: andrea.mathez@unil.ch

S. Narotzky et al. (eds.), *Agricultural Extractivism in the Mediterranean Region*,
https://doi.org/10.1007/978-3-032-05599-6_2

the Strategy Green Generation (GG) to make the agricultural sector a motor for poverty reduction and the growth of its economy. The GMP and the GG are emblematic of these green modernisation schemes (Mathez & Loftus, 2023), based on the 'green economy' paradigm which treats environmental and climatic change not as a crisis but as an opportunity for the creation of new "zones of interaction between capital and various forms of nature" (Bergius & Buseth, 2019, p. 65).

Agriculture is vital to Morocco's economy, food security, and to the lifeworlds of a large segment of the population, contributing 13% to GDP and employing 73.7% of the rural and 38.8% of the total workforce (MAPMDREF, 2018). Yet, Morocco's agriculture faces many challenges, including a negative agricultural trade balance, high vulnerability to climatic variations, and increasing degradation of natural resources—notably water (Taheripour et al., 2020), alongside poverty and precarity among the country's rural population (HCP, 2018). Between 2019 and 2022, Morocco experienced a 32% rainfall deficit—the driest four-year stretch since the 1960s (Bouzidi et al., 2023). In 2025, amid ongoing drought, King Mohammed VI urged citizens to "refrain from sacrificing sheep" for Eid Al Adha, one of the most important days of celebrations for Muslims (Reporterre, 2025). Late spring rains helped replenish dam reserves, prompting a government official to remark, "we have now secured the access to drinking water for the two years to come" (*personal communication*, May 2025). Yet drought, a creeping and unpredictable phenomenon, means even abundant rainfall can obscure the deepening of Morocco's ongoing water crisis. The World Bank group estimates Morocco's water allocation as about 620 m^3 per person per year, corresponding to a relative water scarcity, and which could fall below 500 m^3 by 2030, to a state of absolute water scarcity (WB, 2023). According to a group of Moroccan experts, this state may already have been reached (Aboufirass et al., 2022). While Moroccan media prefer the term *water stress* to emphasise climate factors, some Moroccan researchers talk of *water scarcity* or *distress*, to guide attention to the human factor in this water crisis. As such, Akesbi (2024, p. 1) evocatively frames Morocco's water crisis as the result of the "whims of the sky and irrational choices"[1] made here on Earth. Indeed, Morocco's green modernisation policies did not live up to the supposed "ecologically-sound replay" of the first green revolution (Patel, 2012,

[1] Unless specified otherwise, all quotes are translated by the author from either French or Arabic to English.

p. 37): The agricultural sector now consumes 87% of the water resources (HCP, 2020) and depletes aquifers by a billion litres annually (Himmich & Hatimy, 2023), amplifying Morocco's agricultural and general water crisis.

Against this backdrop, the chapter sheds light on farmers engaged in alternative agricultures—agroecological farming at the margins of agribusiness and these modernisation policies, contributing to the (re)localisation of food and the (re)communalisation of social life (Mathez, 2024)—and explores how they struggle for water in Morocco's increasingly thirsty landscapes. Drawing on several years of ethnographic research on alternative agricultures, the aim is to sketch out a panorama of subsistence and petty commodity producers and their strategies to maintain their livelihoods in different geographies, from the outskirts of Morocco's big cities to the oases and mountains. To do so, I first outline the larger socio-ecological processes in which alternative agricultures are embedded and with which they have to compose.[2]

Morocco's Green Agricultural Policies

The GMP aimed to transform agriculture into a specialised, standardised agro-industry producing 'high-added-value' goods to drive growth and reduce rural poverty. The reliance on foreign capital and the need for new avenues for profitable investment, led the state to grant 700,000 hectares of agricultural land through public–private partnerships to both national and foreign actors with substantial social and financial capital, thereby favouring large-scale capitalist over smallholder family farming (Mahdi, 2014, 2020). This shift reflects a broader political economy shaped by *philanthrocapitalism*, that is, a context in which business leaders and corporate interests have increasingly influenced international development programs, driven by the belief that private enterprise and market forces are the primary engines of 'social progress', and that business expertise outweighs other forms of knowledge (Moseley, 2017). It partly also reproduces colonial-era territorial divides: under the French Protectorate (1912–1956), investments were directed to the 'useful' agricultural zones, marginalising mountainous and arid regions (Jouve, 2002; Kadiri & ElFarah, 2013). Post-independence policies continued this model, channelling public funds—largely financed by foreign loans—into large-scale

[2] For an in-depth analysis of the GMP see Akesbi (2012) and Mathez and Loftus (2023).

hydraulic infrastructure and irrigation schemes and intensive agricultural exploitations in these same zones. The construction of massive dams and the development downstream of irrigated areas were favoured to serve large-scale export-oriented cultures to generate foreign currency for debt repayment, entrenching Morocco's *agro-export model* (Akesbi, 2024). Rather than breaking this cycle, the GMP and GG have intensified it, based on localised irrigation infrastructure in the form of the triptych of drip irrigation, water pumps, and solar power.

In the GMP and GG, the planting of trees and the extension of drip irrigation have been presented as environmentally sustainable actions in themselves, although they have contributed to producing thirsty cash crops for export and the overexploitation of the groundwater tables (Mathez & Loftus, 2023). The environmental impacts and risks associated with the rise of solar-powered irrigation have been largely underestimated (Ferchichi et al., 2024). Driven also by individual initiatives—since farmers perceive solar energy as inexpensive compared to the rising costs of other energy sources—its use in agriculture has expanded significantly (ibid.). Under the Agricultural Development Fund, localised irrigation systems are subsidised up to 100% for farms smaller than 5 hectares and for collective projects, and up to 80% for larger farms. However, equipment related to solar energy remains excluded from these subsidies and is thus treated as a private investment. Although the regulatory framework for solar pumping in irrigation is relatively well developed, its implementation remains limited, and farmers receive little guidance on how to properly size installations (ibid.). In a context characterised by sectoral fragmentation and weak coordination among institutions responsible for energy, agriculture, and water management—and sometimes even contradictory state effects, as discussed below—the combination of generous subsidies for drip irrigation and the relatively low cost of solar energy makes this irrigation triptych an effective means for agricultural intensification.

King Mohamed VI's (2018) speech raising concerns about rural employment, followed by his call to revise the GMP, opened the space for broader criticism. In response to it, the GG,[3] launched in 2020, gives "the

[3] Interestingly, while the 'Green Morocco Plan' is simply the English translation of the French term, the strategy GG is commonly called 'Génération Green' with 'green' said in English, as if with a change of language, one can revive the promise associated with this word. In the First Green Revolution, *green* stood for high-yield agricultural techniques, implicitly opposed to *red* communist social reforms, in the New Green Revolution, it stood for its ecologically sound reply (Daño, 2007).

priority to the human element" and to "the perpetuation of agricultural development" (MAPMDREF, 2020, p. 2). Discursively, the GG appears to acknowledge one key shortcoming of the GMP: the neglect of smallholder family farming, which remains central to Morocco's social stability and food security and proved vital during the Covid-19 crisis, when returning urban workers were sustained by rural households (Kadiri *in ElMazouni*, 2020; Sraïri, 2021). Yet, the GG does not indicate a change of direction regarding the second shortcoming, its natural resource exploitation (Akesbi, 2024). Subsidies for the irrigation triptych have remained in place, facilitating groundwater extraction and sustaining agricultural 'development' and the export of water for 'higher-added-value', notably in the form of citrus fruits, strawberries, watermelons, and to a lesser degree avocados or dates. These plants are not inherently 'thirsty' but have become so through intensified production and expansion into arid regions. As such, in 2007, an agricultural 'boom' around watermelon cultivation began in the Zagora region, driven both by local farmers and investors on former pastoral lands surrounding the ancient oases. This expansion, enabled by individual access to groundwater and supported by the GMP and GG, led to a drinking water shortage for the local population. Following social unrest in 2017, the Ministry of the Interior intervened to restore social stability and secure drinking water for Zagora and surrounding areas. This episode illustrates the state's conflicting roles and priorities: while measures were introduced to regulate groundwater use to secure drinking water, the state's approach to agricultural water use remained ambiguous (Bossenbroek et al., 2023). The Ministry of Agriculture had actively promoted agricultural expansion in the valley, including watermelon farming, through subsidies for irrigation basins and drip systems, accepting applications without formal authorisation from the Water Basin Agency (ibid.). It was only after the mounting public outcry and an extreme drought in 2022 that the Ministry of Finance cut the subsidies for drip irrigation used for the production of watermelon, avocado, and citrus fruits (Agrimaroc, 2022).

However, even in the absence of subsidies, the current irrigation infrastructure enables ongoing pressure on Morocco's water resources. Moreover, as illustrated below, actors involved in agribusiness and philanthrocapitalism can afford to extract resources and subsequently move further, showing little concern for sustainable resource management.

> The Garden[4] is presented as a philanthropic and agroecological initiative, managed by a French family that made its fortune in the agro-industry. It appears as an oasis, standing in stark contrast to the desolate landscape of abandoned and dry tree plantations lining the road we drove along on this hot September day to reach it. The manager running the place presents himself as a "farmer's son" saying "nature is in our blood". Trained as a golf landscaper and a former golf player, he has travelled all over the world and now commutes 120 kilometres back and forth each day between his home and the project site. He explains: "Here, we're lucky, we are in a water-rich area"—contrasting with what we just saw. I ask him if he uses water-saving practices to what he replies "I'm not going to bother saving water. Here, we have water from three wells and when we don't have anymore, that's how it is. People come here to eat healthy food with the flavours of Morocco. We harvest for the hotel; vegetables harvested at noon or in the evening have a different taste—here, it's really fresh". He hands us some herbs which all have an incredible taste and smell and further explains: "The business model of this place is a hotel on-site for international multibillionaires, and on the other days, for the wealthy of Morocco. What keeps me here is the social aspect—the idea that it's an open domain. The estate is divided into a hotel section and a social section with workshops for children".

In 2022, in his intervention in the parliament on the national water policy, M.A. Akhannouch, the head of government, and former agricultural minister under which these green plans were launched, called for a collective responsibility in the management of the water resources: "Just as water is a right for all, preserving and protecting it from waste is a responsibility we all bear, each from our own position" (Head of Government, 2022, p. 16). Yet, as with the climate crisis and capitalist endeavour, it is not those most responsible who bear most of the consequences. While these *green* policies mainly benefit large-scale and export-oriented farming, they also co-produce a margin in which farmers engaged in alternative agricultures increasingly struggle for water to sustain their livelihoods.

[4] Anonymised project in the Haouz of Marrakech.

The Struggle for Water at its Margins

In what follows, I outline a range of alternative agricultural experiences, encompassing both their 'loud' and 'quiet' expressions. In Mathez (2024), I define *loud alternativity* as practiced by farmers engaged in subsistence or petty commodity production who explicitly align with movements like agroecology and/or position themselves in opposition to dominant agricultural models. *Quiet alternativity*, by contrast, refers to everyday agroecological practices with positive socio-ecological effects that farmers do not necessarily frame as alternative or disruptive—whether due to discretion or lack of awareness. I begin by exploring examples of loud alternativity near Morocco's major cities, before turning to quieter forms in the Atlas Mountains and the Figuig oasis.

Some Recent Agroecological Sprouts Flourished and Then Vanished

Although Moroccan family farming—historically based on subsistence and petty commodity production—has long relied on ecological synergies, farmers are increasingly turning to synthetic fertilisers and pesticides. During my fieldwork in the regions of Rabat and Marrakech, I visited various *souqs*—traditional open rural markets—asking farmers and vendors whether they knew anyone growing *bla dwa* (without chemical inputs). Very few did. *Loud alternativity*—agroecological initiatives focused on producing and marketing food *bla dwa*—is relatively recent. Though some early roots may predate them, such initiatives began to sprout in the early 2000s, spurred by food safety scandals, growing awareness that *beldi*[5] does not mean organic, and concerns over the health impacts of agrochemical residues. These factors, combined with the personal commitments of early pioneers and an emerging market for agroecological products, led to the rise of local initiatives across Morocco. These grassroots initiatives stem from a wide range of actors: farmers, retirees, returnees from the diaspora, urban consumers, young activists, civil society groups, and local associations—recently also supported by research and development projects, especially in oases. Lacking formal public support, they have, nonetheless, formed informal networks over the past two decades, communicating via events, WhatsApp and Facebook groups. The

[5] *Beldi* literally means "of the country" and is used to describe anything considered local, Moroccan, traditional, or authentical (Rachik, 1997).

strongest growth has occurred around Rabat-Salé, Casablanca, and Marrakech, where urban demand and purchasing power are concentrated. There, agroecological efforts have begun to institutionalise, notably with the creation of the Network of Agroecological Initiatives in Morocco (RIAM) in 2009, and the adoption of Participatory Guarantee Systems (PGS) as a low-cost alternative to formal organic certification.[6]

One of these very first agroecological initiatives has grown in Dar Bouazza, a neighbourhood at the outskirts of Casablanca, kicked off by Fettouma Djerrari Benabdenbi who presents herself as "an early influencer and trigger of ideas". A sociologist by training, she has been working since the 1980s in the domain of social economy and advocating for rendering visible women's entrepreneurship. In 2001, at a conference in France, she met Pierre Rhabi[7] and subsequently invited him to Morocco. She picked him up at the airport, and in the car looking at the landscape passing by, he said "Fettouma, your country is in the process of desertification". Fettouma repeatedly says that this sentence changed her life, and generated the idea of creating a centre for agroecology. She wanted to become a farmer herself, so that she would not only be preaching words. In the same year, she quit her job in a social security company and bought a tiny plot of land in Dar Bouazza, which she would later name *Jnan el Kbir*, The Big Garden. In the following years, together with Pierre Rhabi and others, they provided training and did much advocacy work to create both the supply and demand for agroecological products. In 2005, they founded the association *Terre et Humanisme Maroc* (*THM*) and in 2006, they set up with several farmers and consumers of the neighbourhood a community-supported agricultural system producing around 60 agroecological baskets a week.

While the *Jhah el Kbir* and THM still exist, the agroecological dynamic in Dar Bouazza after flourishing for a few years subsequently vanished due to increasing water scarcity and to the real estate market. Dar Bouazza has become a posh suburb of Casablanca, slowly replacing Bouskura in the race to attract the affluent citizens of Casablanca. This made land prices spike, motivating many farmers to sell their land, and among those who stayed, the lack of rainfall combined with decreasing ground water levels forced many producers who could not afford to dig ever deeper wells to

[6] cf. LeMeilleur et al. (2019), LeMeilleur and Sermage (2020)

[7] An Algerian French farmer, writer, and thinker who become a popular figure in French-speaking agroecological networks.

abandon agriculture. One of them now runs a *Hanout*, a small grocery store, in Dar Bouazza and some others have become taxi drivers in Casablanca. The groundwater levels are said to have decreased from 11 m to 120 m below the surface today, and since Dar Bouazza is close to the sea, some wells now touch on salt water, contributing thereby to the salinisation of the soil.

Drought creates and reinforces existing inequality, since, contrary to rather affluent persons, some farmers do not have the resources to invest in another plot of land or to continue drilling even deeper wells, as explored below. Overall, the larger the farm and the greater the farmer's financial capital, the more likely they are to access groundwater (Er-Rayhany et al., 2022).

From Rainfed to Irrigated Agricultures: Changing Consumption Trends and Peasant Agroecology in the Hinterland of Rabat

Although operating on a much smaller scale, smallholder family farming has also drawn criticism for contributing to water stress through unsustainable irrigation practices and a shift towards more water-demanding crops. As noted, generous subsidies have enabled widespread access to groundwater—covering part of drilling costs and making irrigation energy affordable via butane or solar power. This has spurred the rapid spread of so-called 'private irrigation', including pump installations along rivers and aquifers, both within and beyond state-developed irrigation zones.

Despite a relatively comprehensive regulatory framework introduced in 2015 for solar-powered irrigation systems—from import standards to on-farm use—key implementation decrees remain pending (Ferchichi et al., 2024). Moreover, while subsidies for drip irrigation require permits from the Hydraulic Basin Agency, enforcement is often undermined by corruption, false declarations (Molle & Tanouti, 2017), and institutional leniency (Collas, 2022). Authorities deliberately apply a laissez-faire approach of continued groundwater use in light of its economic benefits and the resulting social peace (Bossenbroek et al., 2023). According to the Minister of Equipment and Water, of the 372,000 wells registered nationwide, 90% remain unauthorised (L'Économiste, 2023). The following experience of this agroecological farm, run by Mohamed and Irene[8] in the surrounding of Rabat echoes some of these practices:

[8] In accordance with them, I cite their first names only.

> Mohamed explained that while the GMP would have made available subventions for drip irrigation infrastructure, but that "just thinking of writing an administrative letter, makes me shiver"—and without these subventions, the solar panels are too expensive. So, Mohamed keeps running his old gas-oil pump and stays away from home three nights a week to irrigate some hours a night his trees and vegetables on a plot of land further away. Irene, his wife, tells me that they should have got a subsidy of 300,000 MAD (a bit less than 30,000 euro) for their wells, but when everything was ready, they were asked to pay a *Bakchich*. They refused to pay the bribe out of principle—and since then they have been struggling to pay this money back, disgusted by the administrative procedures and practices. As such, when their car was stolen, they did not even report it to the police.

Mohamed grows everything he can find, which leads him to plant not only what sells well but also some varieties for pleasure, for the aesthetics or just for the sake of diversity. Over the years, Mohamed and Irene have surrounded themselves with an affluent but also a very demanding customer base. With a huge knowhow, long labour-intensive days, and collaborations with other farmers, they manage one of the most diverse agroecological farm I have seen in Morocco to offer their clients the year through a vast variety of products (Fig. 1). Yet, some clients are frustrated

Fig. 1 Mohamed running his gas pump. Credit: Andrea Mathez, autumn 2022

when they do not find tomatoes, avocados, and other fruits and vegetables now consumed throughout the whole year. In a humorous drawing (Fig. 2) she shared on Facebook, Irene sought to raise her clients' awareness of seasonal produce. It shows Mohamed listing what's available, while clients express frustration over the absence of apples, eggplants, or Chinese cabbage (see below).

In the winter of 2024, some trees died, such as the avocado and banana trees, due to the prolonged drought and the harsh wind and Mohamed said: "I irrigated with my tears". However, seeing it as his responsibility to provide healthy food tailored to his clients' specific requests, he would, for instance, secretly water the celery at night, out of sight of his employee—whom he considers an adoptive son—because the latter believed it used too much water. He further comments that

> Organic farming is a bad passion, it's like a husband who takes drugs, who drinks, who gives you less than what you give him, but you still love him. I can work long hours, but the lack of water makes me feel my back and my age.

Fig. 2 Drawing by Irene raising awareness about seasonal products. Credit: Irene

The labour-intensive work and high-quality food and service that have attracted many affluent clients—and now sustain the farm—are becoming increasingly difficult to maintain. Mohamed, whose parents were agro-pastoralists, says his true passion lies in aromatic herbs, trees, and livestock; he grows vegetables mainly to meet customer demand. Looking ahead, he envisions shifting towards a sylvo-agro-pastoral system, cultivating vegetables for just 10 to 15 households, as

> without the vegetables, there would be enough water, even in the future, because if you water the trees properly, they don't require too much water.

Unlike Mohamed's meticulous management of scarce and costly water resources, Chaudier (2022) notes that farmers who received fully subsidised irrigation systems often lacked incentive to conserve water, having made little personal investment. Similarly, Ferchichi et al. (2024) highlight widespread over-irrigation in solar-powered systems, where water use often exceeds crop needs and monitoring tools are rarely employed. This is largely due to the perception of solar energy as 'free', given its negligible marginal cost, and farmers tend to irrigate during daylight hours, lacking battery storage, which leads to inefficient and suboptimal irrigation practices (ibid.). Agroecological initiatives are not exempt from these issues. Grippon (2018), for instance, observes increased water use among projects near Rabat, many of which rely on year-round vegetable production—more water-intensive than traditional agropastoral systems.

However, many peasant farmers who co-founded the FermesBioShoul cooperative to market their agroecological produce told me they can no longer sustain their families using their parents' farming techniques. Declining rainfall has reduced yields of rain-fed winter crops, cereals, legumes, and animal fodder and thus also the number of animals they can raise. Although spring 2024 saw an exceptional bloom, Mustapha Belharcha, a co-founder, noted that such events have become rare. In recent years, he said, conditions have been much drier than in his childhood. During our many encounters, he pointed out plots of land that once held ephemeral pools teeming with birds that remain now dry at this time of year.

The semi-subsistence agroecological farms I visited in Morocco, though often focused on vegetable sales, still rely on agropastoral systems—growing cereals for household use and animal fodder. However, subsidies for irrigation equipment, whether from the state or international donors,[9] have encouraged a shift towards fully irrigation-dependent crops, such as certain fruit trees and vegetables. This shift is also tied to broader changes in consumption patterns, including rising demand for fresh produce and the growing expectation of year-round availability (Bossenbroek et al., 2023). As Hassan Drabli, a farmer in Shoul, explains:

> My parents didn't have a well, so they only grew cereals and legumes, and they did some vegetable farming on a small plot during the winter for their own consumption. Each vegetable had its season, and we didn't eat everything all the time. Now, we can eat watermelons almost all year long, and that's not natural. [...] I know that people ate legumes in their season and ate a lot of what grew spontaneously, like artichokes and onions. Even though I'm not very old, I remember that during the onion season, we only ate onions in tagine and nothing else—it was actually very good. [...] Even meat, we ate once a week, on the day we went to the *souq*. There was no electricity or refrigerator, we bought what we would consume that day, and sometimes we bought a little extra to eat part of it and dry the rest. We didn't consume much meat; now, people eat a lot of *rumi*[10] chicken. Before, there were only *beldi* chicken; and when a guest came to our house, we had to run after a chicken to cook it [He laughs].

Hassan also raises his concerns about a neighbouring foreign-held farm, which had drilled 80 wells for avocado production. In the broader international context, where avocados are promoted as a healthy, plant-based superfood (Wilson, 2020), the GMP has contributed to the expansion of avocado cultivation in Morocco, aimed at both domestic consumption and export to European and Middle Eastern countries. Avocados, once considered a seasonal luxury, are now a common served drink (either mixed with orange juice or milk) all year round in *Mahlabas*[11] and cafes in Morocco's cities.

[9] Some farmers of the cooperative FermesBioShoul got their solar panels or drip irrigation through a partnership with Fairtrade International.

[10] *Roumi* refers here to industrial and imported breeds of chicken, and, in opposition to *beldi*, generally to anything 'modern' originating from the West (Rachik, 1997).

[11] Literally, a dairy, it designates a place selling breakfast and snacks.

Navigating the "whims of the sky and irrational choices"[12] *in the Atlas Mountains*

Despite the Moroccan state's incessant efforts to 'modernise' its agriculture and farmers (Mathez & Loftus, 2023), there remain some *quiet alternative* agricultures at the margin of Morocco's green policies who continue to farm according to their own aspirations. Like Mena Abaragh,[13] a peasant farmer in the Haouz of Marrakech, who follows traditional agroecological practices, and laughingly denies when asked if she had already heard of ecological agriculture. Yet, she instinctively expresses a kind of agroecological spirit herself.

> "There is nothing equal to agriculture; it encompasses all that is good. Everyone can eat, the birds, ants, and people. Its benefits never end". However, she notes that all these benefits hinge entirely on water availability. "In the past, rainfall was more abundant, and despite the hard work, harvests were plentiful and joyful, supported by animals during threshing". Today, while machines ease ploughing and harvesting, the rain has diminished. "We used to have cows—where are they now? Where are the animals?" she asks. "Drought has changed everything. [...] Everything depends on water. We have no solution—only prayer".

Mena continues to run the family farm, but her children have left the village—a place lacking basic public services like schools and healthcare, and where water is increasingly scarce. This highlights the consequences of a sectoral agricultural approach pursued at the expense of integrated rural development and public service provision.

Contrary to Mena, Boujemaa Afilal, a peasant farmer in the Atlas Mountains near the Yagour Plateau, actively experiments with new crops and techniques to sustain his family's livelihood. After his father's death, his mother Zahara was forced out of their home by relatives and, unable to raise all her children alone, sent him to live with an uncle in Kenitra. There, he first learned to read and calculate in a *hanout*[14] and later worked as a cook in Casablanca, where he was even offered to take over a café.

[12] Akesbi (2024).

[13] These quotes are from fieldwork conducted jointly with Fatiha ElJazouli, who also translated them from Amazigh to French. This conversation was also filmed by Hicham Idrissi Semlali, and parts of it can be seen in the documentary, *Everyone does their part*, by Z. Kadiri, F. ElJazouli and H. ElMazouni.

[14] A small local grocery store.

Still, he always wanted to return to his village, recalling his father's advice: "Don't give up the land". Today, farming alone would not provide food and financial self-sufficiency. Instead of abandoning agriculture, Boujemaa engages in a pluri-activity characteristic of rural livelihood strategies (Kadiri et al., 2015; Mahdi, 2016): he works as a tourist guide while his mother gathers firewood and tends animals, and his wife, Karima, cares for children and manages household tasks, including preparing meals for family and guests.

In their and surrounding villages, around 6000 walnut trees were planted under a GMP project, but when the engineers in charge of the execution of the project (cf. Mathez & Loftus, 2023) failed to return, most farmers—unwilling to use scarce water on a project they did not feel ownership of—let the trees die. Boujemaa did not only irrigate his trees, but he also researched suitable apple varieties for this altitude and planted 300 Gala apple trees along with around 100 pear, cherry, and walnut trees. Despite this diversification, he notes that, for income, he mainly counts on the apple orchard destinated for the regional market, and maybe on the saffron, which he has started to cultivate as the first one in the region. Boujemaa irrigates according to water availability, aiming not for immediate harvests but long-term tree resilience, expecting yields five years after planting. In response to declining rainfall, he has adopted drip irrigation, saying traditional methods are no longer viable under current climate conditions.

> The irrigation system I installed is gravitational drip irrigation because I can't afford a pump and the other necessary equipment. I installed the drip irrigation pipes by myself and I dug a pool for the retention of water from the mountains. I find that it is a very practical system for irrigation, much better than the traditional system.

He irrigates with the water coming from the mountains, which is collected in a community reservoir (Fig. 3) and then shared among the families who have a right to water in this area (Fig. 4). The management of the water, a commons, is carried out in a traditional and collective manner, where each member holds a share in the large reservoir that irrigates this area (Fig. 5). From this water reservoir, the irrigation of each one's land is done through a system via *seguia*, water channels (Fig. 6), through which the water flows naturally due to gravity, without pumps or mechanical devices. The community management of the water resources is overseen

Fig. 3 The community-managed reservoir. Credit: Hicham Idrissi Semlali, autumn 2022

by 7 families for a group of 21 families, based on a hereditary principle. In the day-to-day management, there is a principle of trust, but not without control mechanisms. A group of about 10 men from different families are authorised to impose fines. At the beginning of the agricultural year in the spring, a public lottery is held in front of the mosque of the *douar*, village, to determine the order of access to water. Every family has a representative who attends the lottery draw, called *'alane*, where a member of the *jma'a*, community, who can read and write, writes the family names on separate pieces of paper, folding them in such a way that the names are not visible. Then, these slips of papers are given to a young boy to open to ensure no cheating, and the first name drawn will be the first to begin their turn in the irrigation rotation for the year to come.

All members of Boujemaa's extended family have together three days of water per month. When it is his turn for a few minutes every 21 days, he directs the water through the *seguia* into his own retention pool from

Fig. 4 Boujemaa's own water retention pool. Credit: Hicham Idrissi Semlali, autumn 2022

where he can manage it as he wants. He is the only one in the village to have built a pool where he first directs the water to, giving him more flexibility on how and when to irrigate his plots of land.

In the almost absence of public services, Boujemaa's entrepreneurial spirit serves their struggle to maintain their livelihood in the mountains, yet he is probably further away from the "new generation of agricultural entrepreneurs" (MAPMDREF, 2020, p. 11) envisaged under the GG than the owners of the new palm plantations surrounding the oasis of Figuig, as described below.

From the Gardens of the Ancient Oasis to Palm Plantations

Like in other parts of North Africa, in Morocco, irrigated farms are expanding outside the ancient oases on land previously reserved for extensive livestock farming (Er-Rayhany et al., 2022). Within a larger policy of agricultural and collective land liberalisation (Mahdi, 2020; Kadiri & Er-Rayhany, 2019), the GMP and GG favoured these oasis extensions with subsidies to restore or cultivate new palm plantations.

Fig. 5 Boujemaa with the water measurement stick. Credit: Andrea Mathez, spring 2024

In Figuig,[15] the heart of the ancient oasis—with the traditional stratified plantation system, characterised by palm trees, beneath other fruit trees, and on the soil, vegetable, and grasses for livestock—appears increasingly neglected. It is marked by stagnant water in basins, abandoned gardens, and crumbling structures, while the surrounding areas become more dynamic with the expansion of irrigated farms and palm groves. This shift is driven by both endogenous and exogenous reasons, related to processes of capital accumulation that seize upon the economic potential of specific

[15] My gratitude here to Mohammed Aderghal and Amzil Lahoucine from the University Mohamed V in Rabat for inviting me to this interesting field trip in 2022 on which I draw for this section.

Fig. 6 Boujemaa letting the water flow into the seguia. Credit: Andrea Mathez, spring 2024

combinations of soil, water, and solar resources (Molle, 2017). Specifically, in Figuig, there is also an unstable land tenure situation, due to its proximity to the Algerian border and to inheritance disputes, which hinders cultivation and has led to the partial abandonment of the old oasis. The construction of a dam and a new water distribution system, which prioritises newly established farms in the expanded areas over the old oasis, alongside state subsidies for drip irrigation, have further accelerated this transformation. Many of the newly established plantations, 1 to 4 hectares in size, are managed by relatively affluent retirees with limited agricultural experience, often seeking a quieter life and employing both permanent and seasonal workers. They buy land on the outskirts of the oasis and with state subsidies or personal savings establish palm plantations (Fig. 7). Relying on in vitro reproduction techniques—claimed to result in higher yields—these plantations depend entirely on the irrigation triptych

Fig. 7 Palm plantation at the outskirts of Figuig. Credit: Amzil Lahoucine winter 2022

favoured under the GMP and GG. Compared to the few remaining gardens within the old oasis, they are notably less diversified.

In the heart of the old oasis, one peasant farmer we interviewed[16] is also in charge of the water management and distribution via the traditional *seguia* system and uses his technical knowledge and sensitivity to oasis agricultural practices to cultivate family plots according to the traditional stratified system. He mainly grows vegetables for household consumption or for sale during major holidays, avoiding big quantities because consumers often prefer cheaper products from large farms near Oujda over locally produced agroecological goods. Beyond subsistence crops, his most profitable activities are date production—carried out without in vitro methods—and livestock. In his view, successful farming in the oasis depends on expertise in water management, an understanding of the local climate, awareness of consumption patterns, and close collaboration with herders in the surrounding area.

[16] Anonymised

The opposition between farmers in the ancient oasis and those in the newly developed areas does not follow fixed lines; there is a hybridisation of farming models, where traditional models of family farming also integrate capitalist practices, and where continuities emerge through the circulation of knowledge and the mobility of people who maintain ties to both spaces (Er-Rayhany et al., 2022). Still, many people we met, farming in the outskirts of the oasis, aim to insert their production into value chains promoted by the state for regional, national, or international markets. Although many lack structural support to commercialise their production, their investments are aligned with the national strategies for agricultural modernisation and the regionally adapted development plans. These actors probably come closest to embodying the, at best loosely defined, "new generation of agricultural entrepreneurs"—potential new environmental subjects, eager to listen to the logic of the market who might, more easily, be drawn into capitalist relations (Mathez & Loftus, 2023). As an informant of the GG told me in an interview in 2021:

> The goal is to create an agricultural middle class, which is only possible through the creation of value and thus through a value chain approach to increase income. And to commercialize, one must move towards market-oriented agriculture.

Concluding Remarks

Morocco's recent modernisation policies have sought to integrate previously marginalised regions—once dismissed as 'useless Morocco'—into the formal capitalist economy, extending the "frontiers of capitalism" (Harvey, 2014), and thereby intensifying groundwater exploitation. As discussed, the Moroccan state remains central to agricultural transformation via infrastructure and subsidies but also in shaping the broader conditions for economic and socio-cultural reproduction.

A southern Moroccan proverb says "One does not buy the land, but the sky", capturing the long-held wisdom that land without water is worthless. Yet, in 2022, the World Bank's principal economist for Morocco claimed, "It's upon the water policies to adapt to the agricultural policies, not the contrary" (Michbal, 2022, online). Under this paradigm, *green* agricultural strategies risk encouraging overexploitation by those still able to access water with serious socio-political consequences. Hence, as Akesbi (2024) argues, agriculture must be subordinated to water policy, defining production choices for each pedoclimatic zone based on resource availability and domestic consumption needs. This also leads to the uncomfortable question of environmental justice: ***who has the right to farm, and how?*** The experiences of farmers engaged in alternative agricultures—through recent innovations or traditional water management practices, often in varying degrees of interaction with state and other actors—offer a valuable starting point for addressing this question and for resisting forms of "anticipatory ruination" (Paprocki, 2022). Paprocki's anticipatory ruination builds on Naomi Klein's "disaster capitalism" in so far as opportunities for profiting result from destruction, but it differs from the latter as it responds to disasters which have not yet come to pass. By framing places as already doomed, it justifies dispossession and spatial reconfiguration. This process is especially visible in agrarian settings, where "new claims about the death of the peasantry in the time of climate change prefigure the very crisis they anticipate" (ibid., p. 3).

> In the spring of 2024, I happened to be in a car with a Moroccan businessman active in the export of citrus fruits to Europe and the Golf states. While we had completely different political and ideological stances on Morocco's agricultural policies, we got on well, allowing me to question him a bit further to explore if he was not concerned about the water depletion, also related to this kind of agribusiness. He answered my questions optimistically "many intelligent people are working already to find soon enough technical solutions, such as enough desalination facilities to irrigate even the desert". When I interjected that the salt rejected back in the ocean causes other environmental harms to the costal ecosystem, he said: "Anyway, in the long run, we are all dead". He smiled and accelerated his car to a frightening speed while the Soufi song *Ilahana*[17] by the singer Sami Yusuf was blasting out of the speakers:

إلهنا ما أعدلك

O our God, how just You are

مليك كل من ملك

You are the King of all kings

لبيك قد لبيت لك

Here I am, I answer Your call

لبيك إن الحمد لك

Here I am, verily all thanks are due to You

[17] Official translation: https://www.musixmatch.com/lyrics/Sami-Yusuf/Ilahana/translation/english

Acknowledgement My gratitude to Fatiha ElJazouli—without whom our encounters in El-Haouz of Marrakech would not have been possible—for all those hours spent together and to Safae Moudni who taught me a lot about moving around in rural Morocco. My sincere thanks also to Zakaria Kadiri for his helpful feedback on a previous draft of this chapter. My special thanks again to Hicham Idrissi Semlali and Amzil Lahoucine for taking photos during our fieldwork and for allowing me to use them here.

References

Aboufirass, M., Bazza, M., Abdelhafid, D., El Haiba, M., El Harizi, K., Elmeknassi Youssoufi, E., Lamrani, H., Safine, M., & Yacoubi Soussane, M. (2022, janvier 1). Livre blanc sur les ressources en eau au Maroc: Pour une gestion durable assurant la sécurité hydrique du Paysydrique. https://www.academia.edu/105255229/Livre_blanc_sur_les_ressources_en_eau_au_Maroc.

Agrimaroc. (2022, September 30). *Fin de la subvention de l'irrigation localisée pour les cultures gourmandes en eau!* https://www.agrimaroc.ma/pasteque-et-avocat-fin-a-la-subvention-de-lirrigation-localisee/

Akesbi, N. (2012). Une nouvelle stratégie pour l'agriculture marocaine: Le « Plan Maroc Vert ». *New Medit, 2*, 1–12.

Akesbi, N. (2024). L'agriculture en prise avec la crise de l'eau: Entre caprices du ciel et choix irrationnels. *Revue marocaines des sciences politiques et sociales, 35*, 18.

Bergius, M., & Buseth, J. T. (2019). Towards a green modernization development discourse: The new green revolution in Africa. *Journal of Political Ecology, 26*(1), 57–83.

Bossenbroek, L., Ftouhi, H., Kadiri, Z., & Kuper, M. (2023). Watermelons in the desert in Morocco: Struggles around a groundwater commons-in-the-making. *Water Alternatives, 16*(1), 87–107.

Bouzidi, Z., Faysse, N., Mekki, I., Ferchichi, I., Hassenforder, E., & Rinaudo, J.-D. (2023). Gestion durable des ressources en eau souterraine au Maroc et en Tunisie: Quels apports de quelques expériences fonctionnelles pour réfléchir à des solutions locales? *Alternatives Rurales, 9*, 1–23.

Chaudier, J. (2022, juillet 9). Barrages à sec, agriculture assoiffée: Le Maroc gère mal le manque d'eau. *Reporterre, le média de l'écologie - Indépendant et en accès libre*. https://reporterre.net/Barrages-a-sec-agriculture-assoiffee-le-Maroc-gere-mal-le-manque-d-eau

Collas, A. (2022, octobre 29). Au Maroc, l'oued victime des « voleurs d'eau » et de la sécheresse. *Le Monde*. https://www.lemonde.fr/afrique/article/2022/10/29/au-maroc-l-oued-victime-des-voleurs-d-eau-et-de-la-secheresse_6147836_3212.html

Daño, E. C. (2007). *Unmasking the green revolution in Africa: Motives, players and dynamics*. Third World Network.

El Mazouni, H. (2020, juin 28). *Zakaria Kadiri: « Le plat marocain est tributaire du pain et rien ne dit que cela changera demain»*. http://analyz.ma/zakaria-kadiri-dans-les-habitudes-alimentaires-le-plat-du-maroc-est-tributaire-du-pain-et-rien-ne-dit-que-celachangera-demain-il-faudra-donc-agir-en-consequence-et-securiser-au-maximum-ce/

Er-Rayhany, H., Kadiri, Z., Ait Mous, F., & Bossenbroek, L. (2022). The oasis extensions: Typology and heterogeneity of profiles in Toudgha. *New Medit, 21*(05). https://doi.org/10.30682/nm2205m

Ferchichi, I., Kadiri, Z., Taky, A., Mekki, I., Hammani, A., Taouajouti, N., Lacombe, G., Khardi, Y., & Faysse, N. (2024, novembre 6). *Concilier le développement de l'énergie solaire pour l'irrigation et la durabilité des systèmes oasiens.*

Grippon, F. (2018). *Logiques et potentiels de diffusion de l'agroécologie pour une contribution aux enjeux territoriaux de l'arrière-pays immédiat de Rabat, Maroc* [Mémoire de fin d'études]. ISTOM Ecole Supérieure d'Agro-Développement International.; Université Mohammed V; MediTer.

Harvey, D. (2014). *Seventeen contradictions and the end of capitalism*. Profile Books Ltd.

HCP. (2020). Modélisation de la consommation en eau intersectorielle dans l'"économie marocaine, (No. 14; Les Brefs du Plan). https://www.hcp.ma/file/230470/

Head of Government Kingdom of Morocco. (2022). *Réponse du Chef du Gouvernement, M. Aziz Akhannouch aux questions des membres de la Chambre des Représentants relatives à la question centrale: « La politique de l'eau au Maroc »*. (2022, décembre 12). http://www.cg.gov.ma/fr/node/10914

High Commissionner for Planning of Morocco (HCP). (2018). Les indicateurs sociaux du Maroc. [Online]. https://www.hcp.ma/downloads/Indicateurs-sociaux_t11880.html

Himmich, K., & Hatimy, A. (2023). *Le Maroc, un pays bientôt sans eau?* Nechfate. Online: https://nechfate.ma/maroc-un-pays-bientot-sans-eau/

Jouve, A. M. (2002). Cinquante ans d'agriculture marocaine. In P. Blanc (Ed.), *Du Maghreb au Proche-Orient. Les Défis de L'agriculture* (pp. 51–71). L'Harmattan.

Kadiri, Z., & El Farah, F. Z. (2013). Le rural et l'agriculture au Maroc. Entre inégalités territoriales et sociales. Blog Farzyates/Inégalités du Centre Jacques Berques.http://farzyat.cjb.ma/le-rural-etlagriculture-au-maroc-entre-inegalites-territoriales-et-sociales.

Kadiri, Z., & Er-Rayhany, H. (2019). La politique foncière de privatisation des terres collectives à l'épreuve des réalités d'appropriations et des conflits. *Alternatives Rurales, 7*, 111–129.

Kadiri, Z., Tozy, M., & Mahdi, M. (2015). Young fellahs becoming leaders in Morocco. *Cahiers Agricultures, 24*(6), 428–434. https://doi.org/10.1684/agr.2015.0783

King Mohamed VI. (2018). SM le Roi prononce un discours à l'ouverture de la 1-ère session de la 3-ème année législative de la 10-ème législature. https://www.maroc.ma/fr/discours-messages-royaux/discours-royaux/sm-le-roi-prononce-un-discours-louverture-de-la-1-ere-session-de-la-3-eme-annee-legislative-de-la-10.

L'Economiste. (2023, avril 18). 90% des puits répertoriés au Maroc ne sont pas autorisés. *L'Economiste.* https://www.leconomiste.com/flash-infos/90-des-puits-repertories-au-maroc-ne-sont-pas-autorises

Lemeilleur, S., & Sermage, J. (2020). Building a knowledge commons: Evidence from the participatory guarantee system for an agroecology label in Morocco. *International Journal of the Commons, 14*(1), 465–480. https://doi.org/10.5334/ijc.1020

Lemeilleur, S., Sermage, J., & Mellouki, A. (2019). Système Participatif de Garantie pour un label agro- écologique au Maroc. *Alternatives Rurales, 7*, 16.

Mahdi, M. (2014). Devenir du foncier agricole au Maroc. Un cas d'accaparement des terres. *New Médit, 4*, 2–10.

Mahdi, M. (2016). Paysannerie d'autrefois et d'aujourd'hui: Concept et réalité. In M. R. Saïdi & M. Streith (Eds.), *Petites Paysanneries au Nord et au Sud de la Méditerranée* (pp. 1–14). Publisud.

Mahdi, M. (2020). Incidences de l'investissement foncier agricole sur les sociétés rurales. A propos de « la classe moyenne agricole ». Working Paper. https://doi.org/10.13140/RG.2.2.18482.02245

MAPMDREF. (2020). Nouvelle stratégie de développement Agricole, Power-Point presentation (49 slides).

Mathez, A. (2024). Glimpses of embodied utopias, why Moroccan and Swiss farmers engage in alternative agricultures. *Agriculture and Human Values.* https://doi.org/10.1007/s10460-024-10598-9

Mathez, A., & Loftus, A. (2023). Endless modernisation: Power and knowledge in the green Morocco plan. *Environment and Planning E: Nature and Space, 6*(1), 87–112. https://doi.org/10.1177/25148486221101541

Michbal, M. (2022, juillet 25). Stress hydrique: La stratégie « Génération Green » va-t-elle tomber à l'eau ? Entretien avec Professuer Mohamed Taher Sraïri. *Médias24.* https://medias24.com/2022/07/25/stress-hydrique-la-strategie-generation-green-va-t-elle-tomber-a-leau/

Ministry of Agriculture and Maritime Fisheries, Rural Development and Water and Forests* of Morocco (MAPMDREF). (2018). Agriculture en chiffres 2017. [Online]. http://www.agriculture.gov.ma/pages/publications/lagriculture-en-chiffre-2017 *New appellation since 2017.

Molle, F. (2017). Foreword. In E. Lavie & A. Marshall (Eds.), *Oases and globalization: Ruptures and continuities* (pp. V–VI). Springer.

Molle, F., & Tanouti, O. (2017). La micro-irrigation et les ressources en eau au Maroc: Un coûteux malentendu. *Alternatives Rurales, 5*, 1–18. https://doi.org/10.60569/5-A2

Moseley, W. (2017). The new green revolution for Africa: A political ecology critique. *The Brown Journal of World Affairs, 23*(2), 177–190.

Paprocki, K. (2022). Anticipatory ruination. *The Journal of Peasant Studies, 49*(7), 1399–1408. https://doi.org/10.1080/03066150.2022.2113068

Patel, R. (2012). The long green revolution. *The Journal of Peasant Studies, 40*(1), 1–63.

Rachik, H. (1997). Roumi et beldi: Réflexions sur la perception de l'occidental à travers une dichotomie locale. *Égypte/Monde arabe, 30–31*, 293–302. https://doi.org/10.4000/ema.1656

Reporterre. (2025, March 6). Au Maroc, le sacrifice de l'Aïd est annulé, les éleveurs à la peine. *Reporterre*. https://reporterre.net/Au-Maroc-le-sacrifice-de-l-Aid-est-annule-les-eleveurs-a-la-peine

Sraïri, M. T. (2021). Repenser le modèle de développement agricole du Maroc pour l'ère post Covid-19. *Cahiers Agricultures, 30*, 17. https://doi.org/10.1051/cagri/2021003

Taheripour, F., Tyner, W. E., Haqiqi, I., & Sajedinia, E. (2020). *Water scarcity in Morocco: Analysis of key water challenges*. World Bank.

Wilson, B. (avec Lee, A.). (2020). *The way we eat now: Strategies for eating in a world of change*. 4th Estate.

World Bank (WB). (2023). *Maroc. Rapport sur le Climat et le Développement du Pays (CCDR): Note Technique: Penurie d'eau et Secheresses* (Rapport Climat and Développement No. 182262). https://documents1.worldbank.org/curated/en/099052223171010874/pdf/P177376089c18001f08b5c09e715216fca3.pdf

Water, Crisis Governance, and Everyday Experiences of Uneven Stateness in Rural Tunisia

Dhouha Djerbi

Introduction

In October 2022, Tunisia's Ministry of Religious Affairs (MRA) issued a nationwide communiqué, urging all mosques across the country to perform *Salat al-Istisqā'*—a special supplication for rain—calling on worshippers to "pray and beseech God to grant us beneficial rain and to relieve us from drought and hardship" (MRA, 2021). The relief being sought speaks to a drought that has lingered over Tunisia for years, its presence felt since 2015 and interrupted only by brief moments of respite. Since 2017, the drought has tightened its grip; and by 2022, Tunisia recorded a annual precipitation rate of 146.28 mm, compared to the yearly average of 272.24 mm (World Bank, n.d.)—making it the driest in over 70 years (Derbali, 2021). The water levels in Tunisia's dams and

D. Djerbi (✉)
Department of International Relations/Political Science, Geneva Graduate Institute, Geneva, Switzerland
e-mail: Dhouha.djerbi@graduateinstitute.ch

S. Narotzky et al. (eds.), *Agricultural Extractivism in the Mediterranean Region*,
https://doi.org/10.1007/978-3-032-05599-6_3

reservoirs scattered across the Northern, Central, and Cap Bon regions reached historic lows: In the same month that the MRA called for the *salat*, its agricultural counterpart, the Ministry of Agriculture, Hydraulic Resources and Fisheries (MAHRF), reported that the country's largest freshwater reservoir, *Sidi Salem*, was operating at a 17% capacity (MAHRF, 2022).

In the Cap Bon region, I joined *ʿAisha*—a key interlocutor and peasant farmer—on her plot in the summer of 2023 as part of my dissertation fieldwork on smallholder farming livelihoods in post-uprising Tunisia. There, the regional dams of *Chiba*, *Lebna*, *Hma*, and *Bezirk* had been and would remain nearly empty throughout the season. With filling rates averaging just 3%, the scarcity of water was most immediately registered in the unpredictable and intermittent provision of irrigation flows, on which *ʿAisha's* crops depended. The vagaries of irrigation water distribuition managed by the Société d'Exploitation du Canal et des Adductions des Eaux du Nord (SECADNORD)[1] and conveyed via the Medjerda–Cap Bon canal, punctuated our daily conversations, dictated what tasks could be attempted on a given day, and curtailed our ability to plan even modestly ahead, all the while generating a pervasive sense of uncertainty.

Beyond 'Aisha's small farm, the water question reverberated through households, agricultural and water collectives, and state institutions alike—even ministries like the MRA, seemingly removed from direct water governance, found themselves compelled to respond. Tunisia's "water emergency"—a term coined and leveraged by local civil society—eventually prompted a series of policy interventions aimed at rationing and more "efficiently" managing water supplies. Yet, such responses, in an attempt to manage scarcity, put into question the very 'scarcity' of water resources itself; and what emerged as a technocratic mode of crisis governance both exposed and reinforced patterns of ***uneven hydraulic stateness***—wherein the state's presence, authority, and infrastructural reach are differentially articulated across space and population. Through a multi-scalar inquiry into the drought and its responses, this chapter asks: How does water

[1] The SECADENORD is a state-owned enterprise operating under the Ministry of Agriculture. It has the mandate to manage the transfer and pumping installations as well as sell water to major users such as the SONEDE (Tunisia's national water company) and the CRDAs (Regional Commissions for Agricultural Developments) located along the network's route.

governance manifest for those who live under it, and what does it reveal about governance logics embedded in Tunisia's uneven hydraulic order?

Theorizations on *uneven stateness*—where state capacity varies both spatially and functionally—lend conceptual and politcal clarity to the fractures in Tunisia's hydraulic governance, and help, as Heydemann and Lynch (2024) contend, move understandings of state capacity beyond strict binaries of "weak" and "strong" statehood into more generative analyses of stateness as a fragmented terrain of contested practices (Yom, 2024, p. 249). In Tunisia, as across the Middle East and North Africa (MENA) region, state presence has been geographically uneven, privileging more "politically strategic" urban and coastal zones over rural peripheries or "inland" regions. Concomitantly, functional understandings of unevenness in state capacity suggest that states tend to expend and concentrate energy in areas deemed critical for their survival while retreating from domains perceived as less consequential to regime continuity (Heydemann, 2024). These twin logics of *resource regionalism* (ibid., p. 45) and territorially selective governance motivate this chapter's understanding of Tunisia's uneven hydraulic stateness and the sub-nationally differentiated experiences of what Anand (2017) terms *hydraulic citizenship*. The latter, in particular, is evident in the persistence of post-2010 popular grievances around access to and control over water—as well as the everyday negotiations with provisioning retreat. Both of these offer a window into how the uneven distribution of state reach is experienced, contested, and rearticulated through the lens of water provisioning.

To make this argument, the chapter draws on existing literature and policy documents to contextualize the history of Tunisia's hydraulic governance within which recent drought-induced rationing measures, and the contentious politics they have provoked, must be situated. Building on this historical-political framing, it then turns to a two-week immersive ethnographic engagement with one peasant household in the Cap Bon region to trace the micro-politics of water access and state presence. Through this dual lens—of structural logics and everyday experience—it becomes possible to examine not only how Tunisia's "water emergency" has been managed from above but also how it is lived and negotiated from below.

Tunisia's Uneven Hydraulic Stateness Through the Eras

Tunisia's hydraulic governance traces its origins to the colonial era, when the French administration concentrated irrigation infrastructure in the most fertile lands to serve colonial accumulation. After independence, President Habib Bourguiba's regime extended these logics, launching large-scale waterworks under three regional "Water Master Plans" (1975–1977) aimed at mobilizing surface and groundwater for agricultural, industrial, and domestic use (Pérennès, 1988, cited in Gharbi & Elloumi, 2023). Oversight was centralized in the Medjerda Valley Development Agency, which reported directly to the presidency in a testament to irrigation's centrality to post-colonial state-building (Bhalla, 2024, p. 5). These infrastructural investments were accompanied by land and agrarian reforms, including the creation of *Périmètres Publics Irrigués* (PPIs) in 1973 and the 1975 Water Code, which formalized water-use rights (Gafrej, 2017, as cited in Gharbi & Elloumi, 2023). Promoted as a means to alleviate rural drudgery, boost yields, and feed a growing urban population, these reforms embedded irrigation within broader state strategies of agricultural modernization and territorial control, rendering rural producers legible to the state and tethering their livelihoods to centralized water regimes. "Direct involvement of the state," Bhalla (2024, p. 9) explains, "became a necessity in all matters of water sector decision-making." These developments laid the institutional and infrastructural groundwork for a mode of hydraulic governance in which access to water was mediated through state-defined priorities of agricultural modernization and productivity enhancement. In turn, the producer-state relationship that emerged during this period forged a form of hydraulic citizenship that was both shaped by—and contingent on—state logics of water mobilization and control (Jouili, 2008; Zussman, 1992). As I show in the sections that follow, these logics continue to shape everyday practices of access and improvisation, even as water sources come under increasing pressure and state priorities shift.

The neoliberal turn of the 1980s reconfigured this model. Under structural adjustment, Tunisia's hydraulic governance shifted from direct provision to market facilitation. Under President Ben Ali, the state embraced private investment as a vehicle for sectoral modernization, earning Tunisia accolades from International Financial Institutions and positioning it as a model pupil of market-oriented reform in the region (Pfeifer, 1999).

Among the most notable reforms was the dissolution of the OMVPI in 1989 and their integration into the Ministry of Agriculture's regional arms, the Commissions Régionales de Développement Agricole (CRDAs), which was framed as a gesture toward liberal-style "decentralization," even as de facto centralized control over water resources persisted (Jouili et al., 2013). The 1993 Investment Incentive Code followed, promoting private engagement in land and water development via targeted subsidies (Elloumi et al., 1991). Rather than representing a total withdrawal of the state, these reforms reconfigured its role: from direct provider to broker of capital, facilitating private-led development while preserving technocratic control over key resources.

Irrigated agriculture remained a cornerstone of Tunisia's hydro-agricultural policy even as the broader orientation of the state shifted towards openness to global markets (Elloumi et al., 1991). PPIs still cover over half of Tunisia's irrigated land, and irrigation infrastructure expansion continues to shape water-sector decision-making (Bhalla, 2024, p. 7). Yet, this expansion increasingly catered to capital-intensive agriculture, buoyed by favorable investment conditions and access to privatized state lands. In contrast, smallholders have faced rising costs and support retrenchment. The removal of irrigation subsidies in the 1990s triggered a threefold increase in water prices (Gana, 1998), while access to agricultural credit dropped sharply: by 2016, state-backed credit recipients were half their mid-1990s numbers (Jouili, 2016, p. 12). These transformations marginalized small and even medium producers, particularly in accessing groundwater. Well-digging permits became a bureaucratic bottleneck. While capitalized actors bypassed restrictions via informal networks, less connected farmers navigated opaque and discretionary state processes (Bhalla, 2024, pp. 12–13). Here, hydraulic citizenship bifurcated: where capital flowed, water followed; where it did not, the state receded in a pattern recognizable across other contexts of water resources privatization (Molle & Mollinga, 2009).

The shifting of national priorities as Tunisia's economy opened to global markets—from state-backed agrarian production to export-led industry and tourism—was also mirrored in national water policy. The 1995 National Water Saving Program introduced subsidies for drip and sprinkler irrigation to manage rising demand from the tourism sector and domestic consumption (Gharbi & Elloumi, 2023). While some small farmers benefited, these subsidies ultimately served a broader strategy of using agricultural exports to fund food imports—what Jouili (2008,

p. 187) calls a "mercantile conception of food self-sufficiency."[2] Meanwhile, public investment in hydraulic infrastructure declined, ultimately undermining the state's contemporaneous ability to respond to prolonged droughts and other climate risks (Lebdi & Maki, 2023). Water transfers increasingly favored coastal and industrial zones, reinforcing a spatially uneven model of governance (Ayeb & Bush, 2019). This spatial and sectoral reallocation, coupled with declining infrastructure renewal, entrenched a geographically uneven governance model. As a result, disparities in water access came to mirror broader regional inequalities in state presence and public service delivery.

These disparities are particularly stark in the interior, where drinking water scarcity persists despite official commitments to principles of national solidarity, or the guarantee of equivalent, even if minimal, service levels for all citizens regardless of geographic location (Touzi et al., 2010, p. 61). Since 1968, potable water has been the responsibility of SONEDE, which formally provides near-universal coverage (SONEDE, 2022). In practice, however, rural access relies heavily on under-resourced *Groupements de Développement Agricole* (GDAs), which lack the capacity for system maintenance and depend on inconsistent state subsidies (AfDB, 2009). Shrinking public investment has led to the atrophy of rural water infrastructure, while higher water tariffs have disproportionately impacted the rural poor while better shielding urban consumers (Touzi et al., 2010, pp. 68–69). These geographically informed and historically rooted institutional divides deepen the spatial unevenness of state provisioning and set the stage for the informal and relational strategies of self-provisioning examined later in this chapter. Indeed, across Tunisia's hydro-marginalized geographies, rural producers and households rely on improvised water access strategies: communal wells, makeshift cisterns, and discretionary irrigation flows. If the post-colonial state once cast hydraulic modernization as a vehicle for rural integration, today's rural poor stand at the edges of those deferred promises, navigating retreat, improvising survival, and negotiating access in an increasingly uneven hydraulic order.

[2] Author's translation from French to English.

Uneven Hydraulic Stateness as Contentious Arena

It is no surprise, then, that the post-uprising wave of popular contention has brought questions of water access and provisioning to the fore even before the full onset of Tunisia's current drought emergency. The contours of these demands reflect and reinforce the geography of uneven hydraulic citizenship, even as the 2014 Constitution formally enshrines water as a basic right. What civil society groups have termed "*struggles against thirst*" (FTDES, 2014) include recurring demonstrations, sit-ins, and road blockades that demand integration into the national water grid and access to the material dividends of democratic transition. Smallholder farmers, in particular, have mobilized around irrigation justice, protesting the unequal distribution of irrigation flows and the rising costs of access, including *saʿat al-inqādh* (emergency irrigation hours) billed at elevated rates during service disruptions. These overt acts of protest are mirrored by more mundane, yet deeply political, forms of labor—most visibly carried out by women and children in rural hinterlands who walk long distances to secure household water. Such quotidian struggles are increasingly framed by civil society actors as evidence of systemic state absence, and leveraged as grounds for renewed calls for infrastructural inclusion and redistributive justice (OTE, 2023b).

In the country's interior regions, where water insecurity is part and parcel of broader histories of underdevelopment, such mobilizations for water often speak to more than mere disruptions in service delivery. They voice deeper grievances over pollution, and resource depletion, issues that have become matters of paramount environmental concern (Arab Barometer, 2024, p. 32). In 2019 for instance, residents of Houaidia—a predominantly agricultural village excluded from SONEDE's water grid—staged weeks-long sit-ins to protest the contamination of their sole spring by a nearby quarry (Nawaat, 2020). Protesters invoked not only the right to clean water, but the village's social and economic fabric, increasingly strained by provisioning failures. Further south, in Gabes, both farmers and non-farmers have long mobilized against the depletion of oasis aquifers by the state-owned *Groupe Chimique Tunisien*. Here, demands for water often draw on agroecological memory and ancestral belonging, standing in tension with the logics of industrial extraction (Robert, 2021, 2023). In both cases, local mobilizations gained resonance through civil society networks that provided media attention, legal support, and a frame that connected hydraulic exclusion to broader struggles over territorial justice.

The post-2011 expansion of civil society has contributed to the emergence of specialized actors capable of translating local grievances into legal and political claims. Among the most prominent is the *Observatoire Tunisien de l'Eau* (OTE), which tracks service disruptions and protest events through its "Carte de la Soif" (Thirst Map), rendering visible what might otherwise be dismissed as isolated or mundane disruptions (OTE, 2023a). Such efforts recast water inaccessibility (as opposed to unavailability) not as a temporary consequence of drought, but as a structural condition of life in hydro-marginalized regions. In doing so, they directly contest technocratic state narratives that frame water shortages solely as environmental misfortunes. Indeed, as precipitation levels declined and drought became a central feature of public discourse, the state increasingly leaned on narratives of natural scarcity to justify service interruptions, with successive Ministers of Agriculture invoking climatic factors to explain water unavailability, helping in the words of Akkaz (2018, p.5) deflect responsibility and obscuring the political character of hydraulic governance. As such, civil society actors have challenged this framing, pointing to its use as a discursive fallback that displaces attention from long-standing institutional failures and infrastructure decay (Houimmel, 2022; Rhili, 2023).

State discourse has, however, not been without contradiction. The 2020 *Plan National de la Sécheresse* acknowledged Tunisia's lack of institutional preparedness for drought, revealing internal recognition of governance failures (Touzi, 2020). Similarly, MAHRF's own 2020 *Rapport sur le Secteur de l'Eau* admitted that water losses in both SONEDE and irrigation networks—estimated at 32% and 40%, respectively—were the result of decaying infrastructure and poor maintenance. The report further criticized the promotion of water-intensive crops such as melons and watermelons in arid zones as symptomatic of inefficient resource use (MAHRF, 2020, p. 20). Despite these admissions, state investment in the water sector continued to contract. In 2021, the Ministry's water budget was slashed from 672 million to 470 million Tunisian dinars (Mbarek, 2023), even as the state reaffirmed its commitment to export-oriented, water-intensive sectors like textiles, agro-industry, and tourism. As Poletti (2025, p. 12) notes, Tunisia's macroeconomic strategy remained geared toward securing foreign currency inflows and addressing external deficits—a strategy underpinned by the World Bank's climate assessment framework and one that subordinated water governance to external fiscal targets. Rather than recalibrating in response to rising hydrological stress, then, Tunisia's water governance adhered to austerity and export-driven logics. The result was a fragmented and reactive set of policy measures, implemented unevenly across regions and accompanied by shifting official discourse oscillating between declarations of crisis and rhetorical assurances. The social reception of these measures, their localized contestation, and their discretionary rollout all underscore the co-constitutive relationship between crisis governance and uneven stateness. In what follows, I examine these tensions in greater detail.

Crisis Governance in an Uneven Hydraulic Order

In Tunisia, where agricultural production has always been tethered to erratic rainfall patterns, drought is both an environmental and political stressor. Recurrent shortfalls in staple crop yields, particularly cereals which are near-totally rain-fed, threaten domestic food supply and strain a state already balancing unsteadily on the tightrope of food security. Further compounding these vulnerabilities is Tunisia's structural dependence on imported grains and its ongoing fiscal crisis, which were exacerbated by global shocks, including COVID-19 pandemic and the Russia-Ukraine war. In 2016, the drought led to a sharp decline in cereal production, leading to an increase in cereal imports the following year. A cascade of other problems followed, including the doubling of fodder prices, a drop in olive oil production, and restrictions of irrigation zones (Lebdi & Maki, 2023, cited in Poletti, 2025, p. 8). Each of these episodes lays bare the fault lines of the country's hydraulic governance, forcing a recalibration of state-society relations around access, control, and distribution. The current drought, the most severe in decades, has reactivated these tensions and triggered a cascade of emergency measures that both expose and reinforce the selective reach of the state central to the chapter's argument about uneven hydraulic stateness.

In response to the threat of another harvest deficit and in anticipation of a growing demand for water consumption in the summer months, the government enacted a suite of emergency measures. These included national water rationing and nighttime shutoffs by SONEDE, irrigation bans for select summer crops, and a moratorium on well-digging below 50 meters. Both reception[3] and enactment of water restrictions are revealing

[3] As Afrobarometer (2024) survey data suggests, most Tunisians view improved water governance—not domestic rationing—as the state's urgent priority.

of the experience of uneven hydraulic stateness, not the least because domestic rationing measures were introduced on already uneven terrain of provisioning. In other words, domestic rationing measures were already part of livelihood strategies of residents who were either disconnected from the SONEDE network or were poorly serviced. But it is perhaps the well-digging ban that was most fiercely rejected, not only for its ecological rationale but because it disrupted longstanding, if informal, modes of rural survival. In areas historically underserved by public infrastructure, illicit wells have long underpinned rural social reproduction, serving as grassroots responses to what is perceived as chronic state in-extension. As the crisis intensified, these forms of self-provisioning—or what were once "survival strategies of the remote rural poor"—spread to urban areas, as city dwellers nationwide pursue reliable water beyond state provision. In the Northeast alone, illegal wells surged from 60,415 in 1980 to 151,850 in 2015, with *'Aisha*'s Cap Bon region reaching an astonishing density of four wells per hectare after the 2010 uprising (Houimmel, 2022, cited in Djerbi 2025).

The ecological rationale for well restrictions including overextraction, declining groundwater quality, and unequal distribution is not in dispute. These concerns are, in fact, acknowledged by producers both large and small-scale I spoke with throughout my fieldwork,[4] even as they persist in illicit well-digging practices. Yet, despite episodic enforcement campaigns, the MAHRF has struggled to curtail the expansion of informal groundwater extraction.

It is not so much that such illicit practices occur at the margins of state power where enforcement is weak, regulatory oversight is absent, or producers operate in spaces beyond the state's administrative reach; nor do they reflect the state's inability to penetrate remote and insular pockets of the countryside. In fact, landed elites have long operated deep wells with the tacit or explicit acquiescence of state officials and the surrounding and often underserviced communities (Houdret, 2012), creating a parallel system that eventually trickled down to smaller producers and rural non-producing households. This long-standing accommodation has ensured the persistence of such practices and their eventual spillover to the less

[4] Interviews with medium-scale commercial producers in Chenini, Gabes; Interview with small-producer in Chenini, Gabes (<5 ha); Interview with commercial producer in Bir Lahfei, Sidi Bouzid, Interview with commercial producer in Sidi Bouzid East; Interview with small producer in Chrifet, Slimane (<5 ha).

fortunate, making recent regulatory attempts that seek to apply restrictions on the latter both politically fraught and practically unenforceable. In effect, the MAHRF's efforts to curb unauthorized well drilling selectively disrupt long-established, if legally dubious, arrangements without addressing the structural inequalities that have made such practices necessary in the first place. What emerges, then, is not an arbitrary regime, but a selective one which—in the testimony of small producers from the Oasis of Chenini, Gabes (Kay, 2023), as well as family farmers from Northern Gafsa (Zaghdoudi, 2023)—penalize subsistence strategies while shielding commercial actors and legacy arrangements.

This reading aligns with Bhalla's (2024, pp. 2–3) work drawing on interviews with water bureaucrats, in which she describes a mode of governance marked by de facto tolerance of legal infractions working hand-in-hand with legal repression and community surveillance. In this configuration, illicit water extraction is "tolerated" as a safety valve, enabling resource users to believe in the "public myth of self-determination" while preserving state control. This "complementary mode of governance"—where state authority is neither wholly absent nor uniformly applied—reflects the dual political logic at the core of Tunisia's drought response. On one hand, the formal prohibition on well-digging signals a national commitment to environmental stewardship and policy coherence. On the other, the tacit toleration of infractions acts as a mechanism of local containment, ensuring social stability while safeguarding elite interests.

The contradictions of Tunisia's hydraulic governance were further dramatized in July 2024 during President Kais Saied's unannounced tour of regional dams. At *Barbara* and *Bouhertma* dams in Jendouba, he observed that water levels appeared sufficient for both irrigation and consumption, dismissed reported shortages as abnormal, and accused unnamed "criminal syndicates" of sabotaging water distribution systems. He repeated these claims at the *Nebhana* dam in Kairouan while acknowledging that the reservoir had not been refurbished since 1969 (Office of the President, 2024). Notwithstanding contradictions with data from MAHRF, whose daily reports show critically low dam-filling rates and the unsuitability of available water for irrigation and domestic use (Bahri 2025; see MAHRF 2024 for dam filling reports on the day of the presidential visit), the performative politics of crisis lie much more in invoking conspiracy to deflect institutional failure through an eclectic crisis rhetoric that has come to embody the water emergency.

This performativity becomes even clearer when considering the president's efforts to distance himself from the unpopular water rationing policies introduced earlier in the year. During the same tour, he declared that "*they say it was the President of the Republic who gave the order to ration water* [...] *except that the President of the Republic never ordered the water to be cut off. The water is available, we see it*" (cited in Ben Hedi, 2024). He followed this denial with an invocation of the 1940s drought—a period of acute scarcity—which, he claimed, did not produce disruptions of the same magnitude as those witnessed today. This analogy served to reinforce his claim that the current crisis is not the product of climatic conditions alone, but of orchestrated, malicious interference. By framing the emergency as an outcome of sabotage, Saied shifted blame away from state institutions, even as those same institutions implemented the very policies now disavowed. The disjuncture between state performance and political spectacle thus materializes the central paradox of uneven stateness: a state that governs selectively, speaks in conflicting tongues, and, in its attempt to reassert authority, ends up revealing the limits and fractures of its own governance. In what follows, I turn to the situated experiences of drought governance in one rural locality in Cap Bon in order to trace how these macro-level dynamics are lived on the ground.

Everyday Experiences of Uneven Hydraulic Stateness

Away from the president's hydraulic spectacle, the contours of state presence look markedly different on a plot just under a hectare where 'Aisha-introduced earlier in the chapter- has farmed her entire life.

The family-shared land, fragmented through inheritance after her father's death, mirrors Tunisia's broader patterns of small-scale family-powered agrarian livelihoods: some siblings have built homes on and rooted lives in their parcels, while others straddle urban and agricultural worlds, returning from Slimane—the urban center nearest to their village—periodically to work their plots. Unwedded, *'Aisha* alone remains in the original family home, its door remains always open to siblings, nieces, and extended family members—a center of gravity for a dispersed kinship network. Whereas her siblings navigate pluriactivity, balancing external wage labor with farm work, *'Aisha*'s days revolve around the rhythms of the land, tending her citrus trees and seasonal vegetables, caring for animals in the family's shared shed, where she keeps a dairy cow and her calf, a donkey and foal, and a small flock of sheep. Beneath the shade of the orange trees, a makeshift chicken coop produces just enough eggs for her breakfast and occasional gifts to visiting sisters. A modest donkey-drawn wagon is *ʿAisha*'s sole means of everyday transportation, which she uses to move around from her farm to her sister's adjacent parcel, helping haul armfuls of wild grasses she weeds from her sister's land to feed her animals. For other inputs, like feed and fertilizer, she relies on her nephews or brothers with cars. But she is otherwise spatially tethered, and for yet another key input, water, she navigates the asymmetries of state provisioning, where public infrastructure does not extend evenly, and access is mediated through informal channels and kinship ties.

Through *ʿAisha*'s account (and, more broadly, that of her extended family), I trace practices of infrastructural improvisation and encounters with fragmented service provision. Embedded in these, I argue, are glimpses into marginal forms of hydraulic citizenship, suspended between formal "entitlements" and the lived reality of contingent access, shaped by geography, class, and gendered hierarchies.

For her irrigation needs, *ʿAisha* pays an annual land size-based subscription fee to local authorities, plus volumetric charges measured by the hour through a water meter installed on her farm. This meter monitors what she calls *mā el-barrage* (dam water)—irrigation water released through her *vana* (the Tunisian Arabic term derived from French *vanne*, meaning irrigation valve)—with periodic technical inspections determining her water bills. Across her farm, a hybrid irrigation system encapsulates Tunisia's decades of agrarian transitions and the tensions therein: modern *goutte-à-goutte* (drip irrigation) pipes serve the olive orchards she collectively cultivates with her sisters, while more traditional basin irrigation feeds the citrus trees and vegetables, the former forming her and her family's primary livelihood in this key citrus-producing region. All water releases follow a schedule set by Nabeul's Regional Agricultural Development Commission (CRDA) and is based on a rotational system wherein users wait in line for their turns to be served, theoretically ensuring equal distribution for all users. The summer of my visit to *ʿAisha*'s farm was extraordinarily hot; and when compounded by the weight of the drought leaving the dam levels in the Cap Bon precariously low the system came under even greater strain. As a result, *ʿAisha* endured consecutive days without irrigation water.

Irrigation interruptions are never communicated in advance, but for *ʿAisha*, they can be deduced through a mix of pragmatism and politics. Her ability to read these signals is rooted in years of navigating the opaque rhythms of the distribution system, and in a keen awareness of her perceived marginal place within it. Although the schedule of water release—or where *'Aisha* and other farmers sit in the queue for water—is supposedly set by the regional authorities, it has never been communicated to her. Instead, she pieces together the sequence of distribution through informal networks. News of water releases spreads through phone calls and word-of-mouth among neighboring farmers and extended family members on neighboring parcels. If, for example, a farmer ten parcels away received water today, *'Aisha* infers that her turn may come tomorrow. When the water is released—again following no particular time during the day—'she receives an alert from neighboring farmers, and she rushes to open her *vana* and irrigate her fields (Fig. 1). Any delay risks missing her window in the rotational system, forfeiting her spot to others in the queue. The "efficiency" of this system depends entirely on whether water is distributed at *regular* intervals and following established rotation logics, allowing

Fig. 1 'Aisha taking apart her vana for an inspection. July 2023. Credit: Author

farmers to anticipate and plan irrigation cycles. During the time I spent with *ʿAisha*, this rarely happened.

The system thus depends less on CRDA planning but on the strength and speed of social coordination. During my visits, water distribution rarely followed a predictable pattern, forcing *ʿAisha* to remain in a constant state of readiness. It also reinforced a second, more politically charged layer of deduction about resource distribution: that she occupies a marginal position in the national hydraulic hierarchy. Put another way, the nature of *Aisha*'s farming, small-scale, destined for local markets and subsistence, renders her low-priority in the water allocation order and precludes her from reliance on institutional assurances for provision. Even under optimal climatic conditions, water distribution has rarely functioned smoothly, and the drought has merely exposed the pre-existing cracks—both metaphorical and literal—in Tunisia's hydraulic infrastructure. In her assessment of the situation, *ʿAisha* explained that "*They [the government] say there is no water, but we all know where the water is going.*" Her testimony speaks to distributional injustice keenly felt by small-scale and family producers cited earlier. But it also reveals the basis of her skepticism towards official discourse on the water crisis, —not so much as a question of technical capacity to deliver, but as a conviction that the neglect is deliberate, rooted in a sustained experience of being structurally sidelined — both materially, through infrastructures that bypass her, and symbolically, through patterns of exclusion that mark her as peripheral. As Jillian Schwedler (2024, p. 235) argues, state capacity should be understood not only through measurable metrics of delivery but also through people's everyday experiences and expectations of what the state *ought* to do. Uneven stateness, to follow Schwedler's understanding, is thus deeply experiential. *ʿAisha* and similarly positioned small-holders register the absence of services as socially and politically meaningful and as a reflection of where, in a national hierarchy of value, they fall.

These perceptions of exclusion cannot be disentangled from global economic logics which couple farmers' competitiveness to their access to productive resources and bind their livelihoods to externally-calibrated trade exigencies. Recent European Union demands for reduced citrus production affecting one of Tunisia's primary export crops resulted in the downgrading of the Cap Bon region within the national water allocation plan, effectively lowering quotas for all regional citrus producers

(Jouili, 2023).[5] At the same time, water rationing measures prohibitting irrigation for "non-strategic crops" —many of which are central to the daily diets of local communities— place an additional strain on the social reproduction of those who, much like *'Aisha*, are, in the eyes of national planning, surplus to requirement.

Even in regions like Cap Bon typically excluded from narratives of underdevelopment and rural discontent associated with the country's interior, —uneven proviosioning fuels tensions between the state's perceived ability to deliver, on the one hand, and its willingness to act in the interest of rural communities seen as marginal to national priorities, on the other. This layered disaffection was made visible one afternoon as Najet—one of *'Aisha's* sisters—walked along the road that connects their family plots to the nearest paved street, the route where she regularly catches the *naql rifi* (rural shared taxi) to and from the city. Years of requests to the municipality to fix the road, which floods and becomes dangerously slick during winter rains, had gone unanswered. Formal petitions were submitted, signatures collected, appeals repeated. But no action followed. In the end, the family pooled their savings and paved portions of the road themselves. When I asked Najet if she had any explanations as to why this was, she put it in the familiar language of bureaucratic indifference toward the "*lives of remote fallāḥīn* [farmers] *in the rif* [countryside]."

In this sense, hydraulic citizenship intersects with broader experiences of infrastructural disenfranchisement and its uneven outcomes are made visible in everyday strategies to fill provisioning gaps, much like the self-financed roadwork described above. Returning to water, service delivery interruptions and delays translated into careful calculations ranging from determining which crops to prioritize, which might tolerate suboptimal irrigation, to which to forgo entirely in anticipation of water restrictions on vegetable irrigation. Each decision entailed careful risk assessment, with daily labor organized around the expectation of deferred water allocations. Among these adaptive strategies, none proves more revealing than the siblings' shared reliance on an inherited shallow well, now situated on a brother's parcel following inheritance division but maintained as a collective resource through a jointly used power generator (Fig. 2). Once a reliable source, decades of seawater intrusion and aquifer overexploitation

[5] I thank Laryssa Chomiak for her helpful prompt to extend the analysis beyond the state to consider the role of global trade relations.

Fig. 2 The well in question. July 2023. Credit: Author

have rendered its water too saline and unsuitable for citrus irrigation and corrosive to both soil and crops.

This seemingly collective arrangement, however, belies deeper gendered inequalities that add an additional dimension to experiences of peripheral hydraulic citizenship that extend beyond spatial and class lines upon which uneven stateness is inscribed. As mentioned earlier with reference to distances women and girls from underserved regions traverse in search for water, the feminization of household social reproduction, including water-related labors, effectively binds gendered identities to the responsibility of securing and managing water at a household level. Such gender-ascribed labors are further intensified by the intimate entanglement of subsistence agricultural production and household reproduction not only in Tunisia (Djerbi, 2024) but elsewhere in Southern agrarian economies (Ossome & Naidu, 2021) where water procurement, small-scale production, and social reproduction form a mutually constitutive triad. In '*Aisha*'s world, this triad takes shape through a stratified system of water-related labor rooted in gendered property regimes. To illustrate, not all siblings rely *equally* on the collective well. *ʿAisha*'s eldest brother, endowed with a larger share of land in accordance with Tunisia's gender-unequal inheritance customs, has dug his own. His move to secure independent water access reflects a pattern of infrastructural individualization that has accelerated in the wake of the state's retreat from supporting smallholder agriculture, while also concealing gendered hierarchies within peasant-family production systems. Though *ʿAisha* never complained to me about the smaller size of her parcel, she did voice frustration about the "separation of generators" between households and, more broadly, about the fragmentation of once-shared resources and with them shared risks. On days when irrigation water is not released, *ʿAisha* and whichever sister comes from Slimane to tend her plot must turn to the compromised well for rescue —a routine that has become a defining feature of their summer work.

Even if the compromised well water now serves only marginal purposes—irrigating hardier vegetables, watering animals, and basic cleaning—accessing this inferior resource demands considerable effort, as the siblings' collective action problem has left the well's pump unrepaired for years. The cost of repair is prohibitive for any sibling to shoulder alone, especially the sisters, whose landholdings and incomes are more limited. The result is a gendered choreography of labor around a deteriorating infrastructure of self-provisioning: moving with practiced coordination, *ʿAisha* and her sister would adjust their rhythm to the sputtering sounds of the pump, responding in real time to the machine's resistance. Bent over rusted machinery, *ʿAisha* offered a running commentary to me as an outsider, identifying the parts of the system, explaining the necessary sequencing, and pointing out subtle (to me) cues in the machine's behavior. The tone is consistently matter-of-fact and the language evidently drawn from years of experience in makeshift hydraulic mechanics. But one morning, as the pump finally gave way and water spilled into the holding basin, her technical explanation gave way to something else. Pausing for a moment, her hands still wet from adjusting the pipe, she remarked: "*This is the life of the rural woman*" [*hayat al-mra al-rifia*], collapsing embodied technical knowledge and the lived reality of feminized water-labor and provisioning failure into the same gesture.[6]

[6] I thank Michelle Weitzel for her most helpful suggestions on bridging ethnographic detail with broader theoretical framing.

Rendered equally precarious by provisioning retreat are a host of within-household labors for which water is an obvious prerequisite. In the domestic sphere, a parallel system of domestic water provisioning through SONEDE is at play. Though *'Aisha*'s household, like those of her siblings nearby, is formally connected to the SONEDE network, state-supplied tap water trickles faintly from her kitchen faucet in a feeble stream that is almost impossible to do any washing with. As a workaround, two provisional techniques are then employed: allowing tap water to drip slowly into containers for later use, or—when sufficient quantities of water are urgently needed—reactivating the collective well through *'Aisha*'s personal generator (separate from the one she uses for on-farm irrigation purposes), incurring energy costs in the process. In addition to these, securing potable water has become an increasingly precarious and monetized undertaking. Like many rural Tunisians whose homes remain only nominally connected or altogether excluded from the national grid, she relies on private, often informal, water vendors who operate in legally ambiguous and infrastructurally peripheral zones (Mzalouwat, 2019). On designated days, a traveling water supplier arrives to fill several large containers that *'Aisha* sets out in advance. This delivery services her household and also several others living on the extended family farm, making the vendor their sole reliable source of drinking water. The alternative—purchasing bottled water from the nearest urban center in Slimane—is both financially exorbitant and logistically complicated for a farmer with limited mobility like *'Aisha*. A few months prior to joining her for my fieldwork, the government had announced a 16% hike in the price of potable water that was justified as a necessary response to the depletion of freshwater reserves after successive years of drought.

These titbits of *'Aisha*'s life from her kitchen to her landholding breathe life into a hydraulic reality on the margins of national water governance plans at a time of acute water stress. They also bring into focus how marginal forms of hydraulic citizenship are enacted in practice. Neither reactive nor novel, coping strategies of managing the consequences of drought management policies draw on an already established repertoire of improvisational labor that has long compensated for opaque water provision and inadequate domestic water supply. While the drought has intensified these coping strategies, adding new labor and monetary costs, they emerge from, and are shaped by, a longer history of fragmented state presence. What might be framed as individualized coping mechanisms of the rural poor are more accurately understood as embedded within broader patterns of governance. Intersecting axes of social differentiation—geographic, classed, and gendered—shape both how and where the state materializes through water infrastructure, and the labor burdens it offloads in doing so. In this light, the everyday labor of securing water does more than sustain life on the margins; it simultaneously conceals and reveals the fragmentation that defines Tunisia's hydraulic stateness.

Conclusions

Tunisia's drought has strained everyday life across the country, laying bare what water justice activists and scholars alike have long warned about: the absence of institutional preparedness for the climate emergency, the unsustainable logics underpinning water exploitation and use, and the lack of a long-term governance strategy that ensures equitable and adequate provisioning for all (Derbali, 2021; Elloumi et al., 1991; Mzalouwat, 2019). In an effort to render Tunisia's hydraulic model better legible to international standards, Tunisia's inability to deliver on "sound" water governance performance indicators has often been interpreted in the language of public sector inefficiency and fiscal mismanagement (see for example, World Bank and AfDB, 2020). Throughout this chapter, I have shown that Tunisia's protracted drought and its resultant water emergency are better understood through the lens of uneven stateness, that is, the differentiated and often fragmented exercise of state authority and capacity across space, time, and social groups.

Here, the state emerges as a shifting assemblage of practices, institutions, and discourses that deliver governance, and withhold it, unevenly (Heydemann & Lynch, 2024). Tunisia's hydraulics regime, shaped by decades of spatially selective development and technocratic planning, bears this logic. It privileges some geographies, crops, and producers while pushing others to the infrastructural margins, thereby fashioning differentiated forms of *hydraulic citizenship* where access to water is both uneven and relationally structured by who is recognized as worthy of state provision. In places like *ʿAisha*'s world, state presence was most acutely registered in its absence: in delayed irrigation schedules, in the anemic flow of household taps, and in the costs of operating private wells and securing potable water. Here, the drought did more than simply reveal state deficiencies in provisioning; it brought to the fore who is recognized, and on what terms, as a water-bearing subject of the state as well as whose labors are conjured in conditions of protracted environmental crisis.

Acknowledgments This chapter builds on the essay, Dams, Decay, and Drought: Waiting for Water in Northeastern Tunisia part of the roundtable on The Politics of Infrastructure in the Middle East and North Africa in Middle East Law and Governance 17(2), 199-271. https://doi.org/10.1163/18763375-20251565.

References

Akkaz, S. B. (2018). أزمة الماء في تونس: العطش، من خطر إلى واقع. Nawaat. [The water crisis in Tunisia: Thirst, from threat to reality]

Anand, N. (2017). *Hydraulic City: Water and the infrastructures of citizenship in Mumbai* (Illustrated edition). Duke University Press Books.

Arab Barometer. (2024). *Aarb Barometer VIII: Tunisia Report.* https://www.arabbarometer.org/wp-content/uploads/ABVIII_Tunisia_Report.pdf

Ayeb, H., & Bush, R. (2019). *Food insecurity and revolution in the Middle East and North Africa: Agrarian questions in Egypt and Tunisia.* Anthem Press. https://www.cambridge.org/core/books/food-insecurity-and-revolution-in-the-middle-east-and-north-africa/074956CEB76745F002560325D1ECB00A

Bahri, I. (2025, July 23). Tunisie: La guerre de Saïed contre la soif. *Kapitalis.* https://kapitalis.com/tunisie/2024/07/23/tunisie-la-guerre-de-saied-contre-la-soif/

Banque Africaine de Development. (2009). *APPUI AU PROGRAMME DE RENFORCEMENT DES GROUPEMENTS DE DEVELOPPEMENT AGRICOLES (GDA) D'EAU POTABLE: REPUBLIQUE TUNISIENNE* [RAPPORT D'EVALUATION]. https://www.afdb.org/fileadmin/uploads/afdb/Documents/Project-and-Operations/Tunisie_-_Appui_au_programme_de_renforcement_des_groupements_de_d%C3%A9veloppement_agricoles__GDA__d_eau_portable_-_Rapport_d%E2%80%99%C3%A9valuation.pdf

Ben Hedi, R. (2024). La vérité sur les barrages visités par Kaïs Saïed. *Business News.* https://www.businessnews.com.tn/la-verite-sur-les-barrages-visites-par-kais-saied,519,139604,3

Bhalla, S. (2024). How water sector reforms institutionalised domination and repression in Tunisia's authoritarian regimes. *Environment and Planning C: Politics and Space.* https://doi.org/10.1177/23996544241285647

CREDIF. (2016). *La Femme, Acteur Emergent: Rapports de Genre et Mutation Multiformes dans le Monde Rural, Etude Sociologique dans le Centre-Ouest de la Tunisie.* CREDIF, Ministry of Women's Affairs, Children and the Elderly.

Derbali, M. (2021, July 5). Le Code des eaux tunisien. Le spectre de la privatisation. *Orient XXI.* https://orientxxi.info/magazine/le-code-des-eaux-tunisien-le-spectre-de-la-privatisation,4898

Djerbi, D. (2024). The social reproductive roots of Agrarian contention: Gendered labor amid peasant struggles in Tunisia. *Antipode*, *n/a*(n/a). https://doi.org/10.1111/anti.13081

Djerbi, D. (2025). Dams, Decay, and Droughts: Waiting for Water in Northeast Tunisia. *Middle East Law and Governance, 17*(2), 199–271.

Elloumi, M., Gara, M., & Lasram, M. (1991). Irrigation et Developpement Agricole: L'Experience Tunisienne. *MEDIT, 2*(1–2), 4–12.

FTDES. (2014). *09. May 2014 Report: Social movements.* https://ftdes.net/ost-newsletter-mai-2014__trashed/

Gana, A. (1998). *Agricultural restructuring, household practices and family farm differentiation: A case study of the region of Zaghouan* [ProQuest Dissertations Publishing]. https://search.proquest.com/docview/304421304?pq-origsite=primo

Gharbi, I., & Elloumi, M. (2023). L'agriculture irriguée en Tunisie: Politiques hydrauliques et politiques de régulation foncière. *Cahiers Agricultures, 32*, 17. https://doi.org/10.1051/cagri/2023010

Heydemann, S. (2024). Seeing the state or why Arab states look the way they do. In S. Heydemann & M. Lynch (Eds.), *Making sense of the Arab State*. University of Michigan Press. https://doi.org/10.3998/mpub.12839265

Heydemann, S., & Lynch, M. (2024). *Making sense of the Arab State*. University of Michigan Press. https://doi.org/10.3998/mpub.12839265

Houdret, A. (2012). *The water connection: Irrigation, water grabbing and politics in Southern Morocco* (SSRN Scholarly Paper No. 2966112). Social Science Research Network. https://papers.ssrn.com/abstract=2966112

Houimmel, M. A. (2022). *The water crisis in Tunisia*. FTDES. https://ftdes.net/rapports/eau2022.pdf

Jouili, M. (2008). *Ajustement structurel, mondialisation et agriculture familiale en Tunisie* [These de doctorat, Montpellier 1]. https://theses.fr/2008MON10021

Jouili, M. (2016). Changements structurels, politiques publiques et petite agriculture familiale en Tunisie. *Transformation des petites exploitations agricoles en Afrique: Le rôle des politiques et de la gouvernance*. Cinquième Conférence de l'Association Africaine des Agro-Economistes, Addis Ababa. https://doi.org/10.22004/ag.econ.249308

Jouili, M. (2023). *Olive oil and water: Moving towards sustainable agricultural trade between the EU and Tunisia* (No. Issue Brief No 1). Transnational Institute.

Jouili, M., Kahouli, I., & Elloumi, M. (2013). Appropriation des ressources hydrauliques et processus d'exclusion dans la région de sidi Bouzid (Tunisie centrale). *Études rurales, 192*(2), 117–134. https://doi.org/10.4000/etudesrurales.9929

Kay, S. (2023, January 30). *Living on the land*. https://africasacountry.com/2023/09/living-on-the-land

Lebdi, F., & Maki, A. (2023). *La sécheresse au Maghreb: Diagnostic, impacts et perspectives pour le renforcement de la résilience du secteur agricole*. United Nations Food and Agriculture Organization. https://doi.org/10.4060/cc7126fr

MAHRF. (2020). *Rapport National sur le Secteur de l'Eau 2020*. Ministry of Agriculutre, Hydraulic Resources, and Fisheries. http://www.onagri.nat.tn/uploads/Etudes/Revue_2020_final.pdf

MAHRF. (2022). *Apports et Lachers des Barrages: Octobre 10 2022*. Ministr of Agriculutre, Hydraulic Resources, and Fisheries. http://www.onagri.nat.tn/uploads/barrages/BARRAGES31-10-2022.pdf

MAHRF. (2024). *Apports et Lachers des Barrages: Juillet 23–24 2024*. Ministr of Agriculutre, Hydraulic Resources, and Fisheries. http://www.onagri.nat.tn/uploads/barrages/BARRAGES-24-7-2024.pdf

Mbarek, G. B. (2023). *How Tunisia's export-focused farming is draining its water*. The National. https://www.thenationalnews.com/weekend/2023/05/05/tunisias-export-focused-agriculture-is-exhausting-its-water/

Molle, F., & Mollinga, P. P. (2009). Hydraulic Bureaucracies and the Hydraulic Mission: Flows of Water, Flows of Power. *Water Alternatives, 2*(3), 328–349.

MRA. (2021). بلاغ صلاة الاستسقاء. [Prayers for Rain Announcement]. Ministry of Religious Affairs. https://www.affaires-religieuses.tn/public/ar/actualites/166977759

Mzalouwat, H. (2019). L'accès à l'eau, une injuste répartition. *Inkyfada*. https://inkyfada.com/fr/2019/03/06/tunisie-eau-potable-chiffres/

Nawaat. (2020, February). طبرقة: أهالي دوار الهوايدية في إعتصام منذ أكثر من 50 يوم. [Tabarka: residents of Houaidia village have been protesting for over 50 Days] *Nawaat*.

Office of the President. (2024, July 22). *President Kais Saied visits several dams to assess the water situation*. Facebook. https://www.facebook.com/photo.php?fbid=889387986553136&set=pb.100064458289062.-2207520000&type=3

Ossome, L., & Naidu, S. (2021). The Agrarian question of gendered labour. In P. Jha, W. Chambati, & L. Ossome (Eds.), *Labour questions in the Global South* (pp. 63–86). Springer. https://doi.org/10.1007/978-981-33-4635-2_4

OTE. (2023a). *Carte de la Soif 2023*. Observatoire Tunisien de L'Eau. https://www.watchwater.tn/fr/blog/2024/1/23/138-carte-de-la-soif-2023/

OTE. (2023b). نقص المياه في أرياف طبرقة: شهادات من الأرض حول معاناة النساء الريفيات. [Water shortages in Rural Tabarka: On-the-ground testimonies of rural women's struggles]. Observatoire Tunisien de L'Eau. https://www.watchwater.tn/ar/blog/2023/10/10/132-nks-lmyh-fy-aryf-tbrk-shhdt-mn-lard-hol-maan-lns-lryfyt/

Pfeifer, K. (1999). How Tunisia, Morocco, Jordan and even Egypt became IMF "Success Stories" in the 1990s. *Middle East Report, 210*, 23. https://doi.org/10.2307/3012499

Poletti, A. (2025). Extractive Water policies as a driver of capitalism expansion in the Mediterranean: Analyzing Tunisia's Commodity Frontier. *Middle East Critique, 1*(21). https://doi.org/10.1080/19436149.2025.2499765

Rhili, H. (2023). The World Bank's water and sanitation policies in Tunisia. Transnational Insitute. https://www.tni.org/en/article/the-world-banks-water-and-sanitation-policies-in-tunisia

Robert, D. (2021). Contestations croisées des nuisances environnementales des industries et des injustices territoriales à Gabès et Kerkennah (Tunisie). *Justice Spatiale = Spatial Justice, 16*, https://www.jssj.org/article/contestations-croisees-des-nuisances-environnementales-des-industries-et-des-injustices-territoriales-a-gabes-et-kerkennah-tunisie/.

Robert, D. (2023). Les réponses socio-territoriales aux mobilisations contre les nuisances industrielles dans la Tunisie post-2011. *La Lettre de l'IRMC : Institut de recherche sur le Maghreb contemporain, 34*, 42.

Schwedler, J. (2024). State Capacity and Contention: A View from Jordan. In S. Heydemann & M. Lynch (Eds.), *Making Sense of the Arab State*. University of Michigan Press. https://doi.org/10.3998/mpub.12839265

SONEDE. (2022). *Stratégie de Desserte du Milieu Rural. Société Nationale d'Exploitation et de Distribution des Eaux*. https://www.sonede.com.tn/strategie/strategie-de-desserte-du-milieu-rural/

Touzi, S. (2020). *Plan National Sècheresse Tunisie*. United Nations Convention to Combat Desertification; Ministry of Agriculture, Hydraulic Resources and Fisheries; Ministry of the Environment.

Touzi, S., Barraqué, B., & Treyer, S. (2010). Le service de l'eau potable en Tunisie:Le modèle de régulation tarifaire face aux défis futurs. *Revue Tiers Monde, 203*(3), 61–80. https://doi.org/10.3917/rtm.203.0061

World Bank. (n.d.). World Bank Climate change knowledge portal. Retrieved June 28, 2025, from https://climateknowledgeportal.worldbank.org/

World Bank, & AfDB. (2020). *Tunisia public expenditure review: A new pact for the transition modernizing the state for better and faster public spending*. World Bank Group. https://openknowledge.worldbank.org/server/api/core/bitstreams/a57cea9a-9223-5730-a305-ce9d9290afa3/content

Yom, S. (2024). Water, stateness, and tribalism in Jordan: The case of the Disi Water conveyance project. In S. Heydemann & M. Lynch (Eds.), *Making sense of the Arab State*. University of Michigan Press. https://doi.org/10.3998/mpub.12839265

Zaghdoudi, M. (2023). *Gafsa Nord: Politique de l'eau de l'État :La soif des petits agriculteurs et de l'eau pour les grandes entreprises et les investisseurs*. Observatoire Tunisien de L'Eau. https://www.watchwater.tn/fr/blog/2023/7/18/129-politique-de-leau-de-gafsa-nord-soif-de-petits-agriculteurs-et-deau-pour-les-grandes-entreprises-et-les-investisseurs/

Zussman, M. (1992). *Development and disenchantment in rural Tunisia: The Bourguiba Years* (1st edn). Westview Press.

Monocultures in Southern Tunisia: Promising Employment, Promoting Desertification and Displacement

Arianna Poletti and Gennaro Avallone

Introduction

In Tunisia, particularly in its southern regions, the pre-colonial status of land meant it was administered as commons. Land grabbing, first by the French protectorate and then by the Tunisian state, led to the fragmentation of collectively owned lands. Initiated in 1986, the IMF's Structural Adjustment Plan, along with subsequent reforms by international financial institutions, contributed to reorienting Tunisia's agricultural production toward exports and increasing its dependence on imported essential

A. Poletti (✉)
Climate Change and Sustainable Development (PhD-SDC), IUSS Pavia (International School for Advanced Studies), Pavia, Italy

Department of Political and Social Studies, University of Salerno, Salerno, Italy
e-mail: arianna.poletti@iusspavia.it

G. Avallone
Department of Political and Social Studies, University of Salerno, Salerno, Italy
e-mail: gavallone@unisa.it

S. Narotzky et al. (eds.), *Agricultural Extractivism in the Mediterranean Region*,
https://doi.org/10.1007/978-3-032-05599-6_4

goods. Tunisia's proximity to the EU, along with favorable exchange rates and low labor costs, presents a significant comparative advantage for European markets. This dependence on imports, especially during liquidity crises and food shortages, has spurred socio-ecological movements advocating for food sovereignty.

In southern Tunisia, monocultures, often supported by European funds, promise employment but contribute to the disintegration of societies that once relied on the land for sustenance. Specifically, the export-oriented Deglet Nour date monocultures have caused both internal and external population displacement, leaving the female agricultural workforce in precarious structural conditions. By tracing the process of land grabbing in southern Tunisia and applying a political ecology critique, this chapter explores the link between agricultural practices, social movements, and population displacement, focusing on the Deglet Nour date production case study. The chapter opens with a theoretical framework that conceptualizes Tunisia as both a cheapened ecosystem and a "commodity frontier," drawing on political ecology to examine the entanglements of environmental degradation, labor devaluation, and global capitalist expansion. Building on this foundation, the analysis turns to the empirical case study of southern Tunisia, which is developed across three sections: (1) a historical examination of the transformation from communal land tenure to individualized land ownership; (2) a critical analysis of the Deglet Nour date export economy and its implications for agrarian change and landscape fragmentation; and (3) an exploration of the climate migration discourse and its articulation with emerging forms of (im)mobility in the region.

Methodology

This research is based on fieldwork in southern Tunisia conducted as part of a journalistic investigation between 2022 and 2024. The primary ethnographic work was led by Arianna Poletti (Author 1), who lived in Tunisia from 2019 to 2024, enabling sustained engagement with local actors, participation in public events and conferences, and close observation of socio-ecological dynamics related to land grabbing and extractive agriculture. Fieldwork was carried out in collaboration with French journalist Aïda Delpuech, focusing on the environmental and social impacts of export-oriented agriculture. The first research trip in September 2022 covered the governorates of Gafsa and Kebili, including Al-Faouar,

Essabria, Douz, and Jemna. Methods combined semi-structured interviews, participant observation, and direct field engagement. Interviewees included officials at the Rjim Maatoug plantation, local residents, European NGO representatives, and Tunisian civil society organizations such as the association Nakhla in Douz and the El-Jemna date producers' cooperative. Additional research trips in July and November 2023 expanded the inquiry to phosphate-based fertilizer supply chains, with fieldwork in Redeyef, Segdoud, Tozeur, and Nefta, and interviews with local farmers and representatives of organizations such as the FTDES, the Observatoire Tunisien de l'Eau, and Avocats Sans Frontières.

While Arianna Poletti conducted the fieldwork and interviews, Gennaro Avallone (Author 2) contributed to the theoretical framing in order to contextualize the Tunisian case within broader debates on extractivism, land dispossession, and the socio-ecological consequences of global capitalist expansion. Both authors collaborated in interpreting the data and writing this article. To analyze how these issues are framed and represented, this research employs both historical analysis and Critical Discourse Analysis (CDA) to interrogate key concepts such as "desertification" and "displacement." By adopting an interdisciplinary approach—drawing on political ecology, critical migration studies, and discourse analysis—the study challenges reductionist or depoliticized framings of environmental and social change, and emphasizes the importance of methodological and analytical plurality.

The Theoretical Approach: Tunisia as a Cheapened Ecosystem

"Hello, strawberry friends! I love my workplace—delicious strawberries everywhere. And I am picking them for you. A great gift." Gift—it is the term chosen by the Spanish company San Lucar to promote their delocalized production of strawberries on the southern shores of the Mediterranean, specifically in Tunisia. Described by the company as a sunny and warm land during Europe's cold seasons, Tunisia provides the European market with sweet, sun-kissed fruits and vegetables. In this same land, however, it is rare to engage with smiling female workers like the one depicted in the company's advertisement.[1] Instead, along rural roads, female laborers are more likely to be encountered crammed into the

[1] https://www.instagram.com/reel/DEDJeM1tsj3/?igsh=OGI2dHNhY2sxZjg4

pick-up trucks of *caporali* (gangmasters of illegal labor recruitment), earning only a handful of Tunisian dinars per day or per kilo. These workers often have little or no access to the agricultural produce they harvest during the ripening season, nor to the lands they traverse from morning to sunset. Gift—it is the term interrogated by Anna Tsing (2013) in her research on the capitalist value chains and the relationships forged by the subcontracted organization of labor between commodities and humans. In her analysis of the Japanese matsutake mushroom supply chain, the anthropologist writes: "Just as in factories workers are alienated from the things they make, allowing those things to be sold without reference to their makers, so too, things are alienated from the people who make and exchange them. Things become stand-alone objects, to be used or exchanged; they bear no relation to the personal networks in which they are made and deployed" (Tsing, 2017, p. 192). The "quality link" between the smiling worker and her juicy strawberries portrayed in San Lucar's advertisements is unlikely to exist within a monoculture farming system, which mirrors the organization of serial labor in terms of both the detachment from the agricultural product and the exploitative conditions of the workers. Gift—it is also one of the concepts studied by Jason W. Moore in his analysis of the capitalist system through the world-ecology framework. Moore (2016) politicizes the notion of *cheapness*, expanding the understanding of global capitalism beyond mere socio-economic relations to conceptualize it as "a way of organizing nature" (Moore, 2016, p. 6). This formulation aptly captures the logic of monocropping—the mass production of a single agricultural product replicated across vast landscapes. As Tsing (2017) observes, it is within the logic of scalability that the boundary between gift and commodity becomes significant (pp. 77–85). In this sense, monocultures exemplify the historical reconfiguration of socio-ecological relations through the systematic exploitation of both nature and human labor—a dynamic Moore categorizes as Cheap Nature. Building on this critique, Moore also challenges the prevailing narrative of the Anthropocene. He argues that the term reflects a limited, binary worldview that obscures the subordination inherent in extractive ecosystems under capital's expansion. To address this, Moore introduces the concept of the Capitalocene, highlighting how capitalism reshapes both nature and society through its logic of accumulation.

From an anthropological perspective, monoculture scholar Sarah Besky similarly critiques dominant frameworks by positioning the plantation as a metaphor for "the triumph of science and technology over purportedly wild landscapes and people" (Besky, 2022). This underscores how imperial and capitalist expansion has historically reshaped both nature and society. Besky introduces the term Plantationocene to describe "the planetary impact not just of monocrop agriculture, but of a variety of systems of accumulation centered on the production of singular forms of life" (Besky, 2022). She argues that revisiting the plantation as an analytic framework allows for a deeper understanding of social reproduction within systems marked by racialized labor and ecological simplification. In alignment with these perspectives, Davis et al. (2019) emphasize the importance of interrogating the plantation as a space where nature is rationalized and spatially reorganized—processes that are inseparable from the racial politics embedded in these highly structured, disciplined landscapes.

In Tunisia, monocultures dominate much of the landscape, reflecting a country of striking climatic diversity—from Mediterranean coasts to arid desert plains. Despite this varied terrain, an intensive agricultural model is rapidly spreading through orange groves, olive orchards, and date plantations lining the roads connecting the country from north to south. Traveling these routes, one often encounters pickup trucks loaded with agricultural workers precariously balanced in the back. It is their often-invisible labor that sustains the steady flow of fruits and vegetables reaching European supermarkets. Within this context, the relationship between rural communities and their local geographies has been profoundly transformed, reorganizing itself around production chains that impose new internal and external boundaries.

The Case Study: Tunisia as a Commodity Frontier

In this theoretical context, the case study of Tunisia and its southern monocropping practices becomes particularly relevant. Traditionally characterized by small-scale agriculture on collective lands, Tunisia has gradually transformed—beginning with the onset of French colonial occupation in the late nineteenth century—into a patchwork of fields managed and owned by both local and international firms composing the agribusiness supply chain, primarily producing for export—especially to the EU, the country's largest market—rather than for domestic consumption.

As an analysis of crops in southern Tunisia demonstrates, the price of this transformation has been the loss of agricultural diversity and varieties, along with the socio-ecological relationships that depended on this multiplicity, such as collective and mobile land management or traditional irrigation systems, in favor of a single dominant product: Deglet Nour dates, which, alongside olive oil, have come to symbolize the country abroad. This shift has fueled recurrent calls from civil society for a return to food sovereignty, particularly in the aftermath of the 2011 revolution, which challenged not only the political centralization of the Ben Ali regime but also the broader extractive model controlled by a narrow economic oligarchy. This new model is historically rooted in a specific political-economic orientation that has radically altered socio-ecological relations in the South of the country. The new waves of what Marxist scholars describe as the ongoing revitalization of primitive accumulation mechanisms or "new enclosures" (Midnight Notes Collective, 1990)—referring to the dispossession of land from peasantry for profit—are concretely observable in the Tunisian fields, where this research originated through field investigative work. The appropriation of former common lands in the post-colonial period has contributed to integrate Tunisia into the complex network of supply chains connecting the two shores of the Mediterranean through the exchange of imported and exported goods, particularly agricultural products. These connections function not merely as physical routes but as *commodity frontiers*, which Moore defines as strategies of power, profit, and life, crystallizing capitalist dynamics and inter-regional movement patterns (Moore, 2021, p. 2), such as those dividing the Mediterranean. Along these frontiers, a relationship of subalternity between the EU and Tunisia has been constructed and perpetuated. This unequal dynamic is obscured by the rhetoric of a fair, sovereign system of exchanges promoted through international cooperation and investments promising jobs and security. Yet, the system operates within a framework of stark monetary disparity: the EU's strong euro versus Tunisia's weaker dinar. Therefore, Tunisia finds itself among many nations perpetually reliant on foreign currency revenues generated through agricultural exports to sustain the vicious cycle of indebtedness. Consequently, Tunisia represents a compelling case study due to its geographical proximity to the EU market, its role as a border country, and the model of dependency relations it embodies at this historical juncture.

The Deglet Nour date plantations exemplify these processes, establishing a commodity frontier characterized by unequal exchange, the consequences of which have severely damaged Southern Tunisia's fragile ecosystems, exacerbating desertification and displacement. Southern Tunisia is a region characterized by both mobility and immobility, frequently reduced in mainstream media narratives to a "departure hotspot" for Tunisian migrants. A substantial body of literature on migration approaches Tunisian mobility through an anthropocentric lens, framing it as a solvable issue rooted in "underdevelopment" and unemployment—challenges the European Union seeks to address through border control policies and international cooperation programs (Bousnina, 2024). However, migration patterns are rarely understood as a structural consequence of an extractive model that disrupts local equilibrium, as evidenced by the expansion of Deglet Nour plantations and the region's reliance on this monoculture as a primary source of livelihood, often serving as an alternative to smuggling with neighboring countries like Libya and Algeria. Research in Southern Tunisia also facilitates an interrogation of patterns of internal colonization, as described by the Tunisian scholar Sadri Khiari (2003). The liberalization and privatization processes affecting common goods, Khiari argues (Khiari, 2003, p. 98), establish an internal frontier between the northern cities—home to a local oligarchy that has profited from the import-export system—and the rural hinterland. This hinterland is characterized by basins of landless labor, where the post-colonial status of land remains contested between family groups and the Tunisian state, which inherited the land from the French protectorate after independence in 1957. The paradox of desertification driven by dispossession, land appropriation, and lush monocultures becomes even more precarious in the context of severe drought and water scarcity which is affecting the whole Maghreb region, as confirmed by data from reports by international organizations (Lebdi & Maki, 2023). These conditions further undermine the viability of the agricultural model promoted since the 1990s as the sole viable option, a model now increasingly revealed as unsustainable, as the interviewees confirmed.

From Commons to Land Ownership in Tunisia: A Historical Process

In Southern Tunisia, as in many countries with a history of colonization, land ownership remains a deeply political and unresolved issue. It reflects the enduring conflict between traditional systems of land management and ways of life, and the institutionalization of land ownership, which both mirrors and is shaped by the colonial legacy. The history of *Deglet Nour*, the most widely produced date variety in the Saharan region of Tunisia, must be understood within the broader transformation of agricultural practices. This transformation, driven by cycles of dependency initiated during the French protectorate and perpetuated in various forms after 1957, marks a shift from traditional, ecologically complex oasis farming systems to linear palm monocropping. Before examining contemporary agricultural management practices and how this shift has reshaped socio-ecological relations within Tunisia and with foreign partners—both private and public—it is necessary to step back and consider the challenges of land ownership. This perspective situates the Tunisian context within a broader historical transformation of capitalist dynamics that crystallized hegemonic relations between the Global North and the Global South.

"It was once the richest province on the shores of the Mediterranean, called the granary of Rome. It's up to us to do again what the Romans did," wrote Robert Wastelier in 1903, encapsulating the French colonial vision of Tunisia's agricultural potential during the protectorate period (1882–1956). In French colonial discourse, comparisons between Ancient Rome and the European colonial "revival" of Tunisia's so-called "abandoned" lands were common, serving to legitimize the transformation of the "indigenous" landscape under the pretext of population well-being and modernization (Holm & Santmyer, 1964). However, as American agricultural economists Holm and Santmyer (1964, p, 1) observed, "during most of the protectorate period, efforts to increase agricultural productivity were mainly oriented toward the European farmer and brought little direct benefit to the majority of rural Tunisians." Following an initial phase dominated by private and speculative colonial capital, agrarian colonization developed into a systematic integration of Tunisia into the French economy, achieved through the expansion of the land base and the establishment of new customs regulations (Mahjoub, 1987). Under the 1885 "immatriculation" law, European settlers in Tunisia could register the parcels they occupied in French courts and, starting in 1905, began converting them from state, collective, or *habous* (religious) tenure into full private freehold, as they were released from the obligation to pay a perpetual rent to the local administration (Giudice, 2009). This process allowed settlers to become legal owners of some of the most fertile and accessible lands. Meanwhile, many displaced peasants either became underpaid day laborers on these colonial estates or migrated to growing urban centers and mining areas, contributing to both their proletarianization and the expansion of industrial and urban hubs in the country. The resulting agrarian framework restructured the agricultural sector to align with the demands of the colonial state, prioritizing the high productivity of three primary crops—olive oil, vines, and cereals—thereby paving the way for industrial-scale production (Mahjoub, 1987). As Tunisian economist Azzam Mahjoub (1987, nn) observed,

> The centralization of large tracts of land and their redistribution to colonial settlers was the mechanism employed by the colonial state to implement primitive accumulation. This process resulted in the separation of peasants from their essential means of production (the land) and their subsequent impoverishment. On this foundation, colonial accumulation became possible. On the one hand, agrarian colonization began to introduce substantial transformations in the mode of production. On the other hand, and more significantly, the mining sector experienced significant growth, with extractive production accounting for a predominant share of Tunisian exports—over 40% in the 1910s.

This analysis of the reproduction of mechanisms of primitive accumulation in colonial Tunisia links the transformation of the agrarian sector to the development of the mining sector, encapsulating this process within the term "export-oriented agro-mining apparatus." The concept of an apparatus, understood as the organization of an (eco)system, effectively captures the systematic transformation imposed by the imperialist division of labor on the indigenous body—a metaphor for the entire territory and its communities—alongside the early stages of the production chain and the process of relocation. Key symbols of this physical transformation include new French infrastructure, such as railways and ports, which reshaped the Tunisian landscape by pulling it toward the Mediterranean and, by extension, French-European export markets (Mahjoub, 1987). As a result, the colonial period marks the beginning of the delineation of Mediterranean commodity frontiers, which hierarchically separate yet simultaneously connect the northern and southern shores. The redirection of internal trade flows toward external commerce profoundly disrupted the country's socio-economic fabric, as well as its traditional land management practices. Mahjoub (1987, nn) observes: "Through progressive division procedures between fractions, sub-fractions, and families, the so-called *arch*, or 'collective' lands, underwent, particularly in the 1930s, a process of fragmentation and individualization." The immatriculation land system, which effectively favored private property, along with complex and evolving land tenure laws, primarily benefited large landowners and French capitalist enterprises, while the Protectorate accelerated the decline of local urban and rural bourgeoisies who, unable to compete, ceded their properties to European settlers and intermediaries (Giudice, 2009). The emergence of "agrarian individualism" in the North during the 1930s stood in stark contrast to established legal traditions, including Islamic land law. At the same time, colonial land legislation played a crucial role in facilitating the rise of an agrarian bourgeoisie and the expansion of wage labor (Mahjoub, 1987). This transformation of the agricultural sector fundamentally altered socio-ecological relations, laying the groundwork for a prolonged process of proletarianization, exploitation, and class struggle.

While independence initiated a phase of so-called nationalization, it did not fundamentally dismantle the colonial land ownership structures. The nationalization measures of the 1960s primarily targeted lands held by French settlers, yet much of this land was ultimately transferred to state control or acquired by affluent Tunisian actors, rather than being redistributed (Auzary-Schmaltz, 2008). According to Ayeb and Bush (2019, pp. 99–100), one of the key factors that permanently altered the relationship between the peasantry and land during the colonial period was, in fact, the establishment of the cadastral system. The registration was established by the Land Law of July 1, 1885, which served as a model for many territorial legislations passed under French influence. It represented a compromise between several different legal systems, including the Australian model introduced by Great Britain in the colony, the German system, the French Civil Code, as well as Algerian and Belgian laws. This law was later adopted by the legislation of independent Tunisia in 1956 and was subsequently incorporated into the Code de droits réels in 1965. Tunisian land registration, often praised for its greater flexibility compared to other colonial systems, nevertheless reinforced colonial structures that continued to shape land relations well into the post-independence era (Auzary-Schmaltz, 2008). In 1956, Tunisia abolished public *habous*—religious endowments of land or property intended for charitable or religious purposes—and integrated them into state-owned land. A subsequent decree in 1957 dismantled the system of private and mixed *habous*, redistributing approximately one million hectares into the economic sector (Auzary-Schmaltz, 2008). The 1957 law also granted tribal members usufruct rights to collective lands, which could later be converted into individual property. Full ownership was granted for developed plots, while undeveloped plots required a development plan, with failure to meet this requirement leading to the forfeiture of property rights (Auzary-Schmaltz, 2008). This transformation was particularly profound in the southern regions of the country. The shift toward individual ownership significantly altered land tenure and agricultural practices in the south, an area historically characterized by communal ownership and use rights rather than individual titles. In these areas, lands were managed collectively by tribal councils, known as *djema'a* or *majles*, which governed access, seasonal use, and allocation among tribe members based on customary law. This system contrasts with the northern and central regions, where private ownership or *habous* (religiously managed lands) were prevalent. As a result, this transformation not only reshaped land usage but also had a profound impact on the social fabric of communities that had long relied on collective land management practices and transhumant breeding.

The 1960s in Tunisia focused on land nationalization and the establishment of socialist-inspired cooperatives. However, these cooperatives ultimately proved unsuccessful, reinforcing a hierarchical, centralized management structure with cooperative directors appointed by the Secretariat of State (Ben Saad et al., 2010). Membership frequently exceeded the cooperatives' productive capacity, forcing surplus members—many of whom owned land within cooperative boundaries but could not find employment—to lease their land back to the cooperative. This process contributed to smallholder dispossession and increased rural unemployment. Moreover, cooperatives regularly failed to pay land rents and did not fulfill the minimum income guarantees outlined in national development plans, further exacerbating socio-economic challenges in rural areas and paradoxically reinforcing peasant dispossession while empowering large landowners (Ben Saad et al., 2010; Ayeb & Bush, 2019, pp. 106–7). This failure paved the way for a significant policy reorientation in the 1970s, characterized by a pronounced move toward economic liberalization and privatization, largely influenced by international financial institutions and a strategic emphasis on attracting foreign investment. A key turning point in land management came with the 1986 signing of the IMF's Structural Adjustment Program (SAP), which advocated for trade liberalization and widespread privatization. This wave of state disengagement facilitated the rise of a powerful and corrupt oligarchy from Tunisia's major coastal cities, which came to dominate the country's key sectors, including agriculture (Khiari, 2003, p. 96). Tunisian scholar Sadri Khiari summarizes, "the new economic policy has only one aim: to develop exports in order to bring foreign currency" (Khiari, 2003, p. 88). At the same time, a process of social disintegration was unfolding (Khiari, 2003, pp. 98–100). The SAP created a divide between the smallholder family farmers and members of collective land management systems and their access to land, definitively transforming the hinterlands into labor reservoirs of landless workers while promoting the growth of low-value-added manufacturing sectors such as textiles, machinery, and labor-intensive crops like olives and off-season vegetables (Ayeb & Bush, 2019, p. 117).

Domestically, International Financial Institutions (IFIs) increasingly supported the establishment of "duty-free" factories (*sociétés sous douane*), which operated in Tunisia thanks to a series of tax incentives but did not integrate their products into the domestic market (Khiari, 2003, p. 80). This reliance on export-oriented factories, as well as outsourcing and subcontracting, relied—and still relies—on what Gachet (1987, p. 163) referred to as "the only comparative advantage the country possesses": a cheap labor force (Ayeb & Bush, 2019, p. 113).

On a broader Mediterranean scale, the SAP marked the beginning of tariff reductions. The 1995 Association Agreement established a free trade area by removing tariffs on industrial products and gradually opening markets for select agri-food goods with the European Union. Under the 1998 Association Agreement, duties on agricultural goods entering the EU from Tunisia averaged 12%, while Tunisia imposed an average duty of 32% on EU agricultural products (GIZ, 2019). Negotiations for a Deep and Comprehensive Free Trade Area (DCFTA) in 2015 encountered resistance from both authorities and civil society, ultimately resulting in a deadlock. However, the pressure on the country for approval never ceased, and some measures favoring intra-Mediterranean exchanges with the EU bypassed certain blockages nonetheless (Abou, 2021). While the full liberalization of the agricultural sector remains a significant concern for Tunisian authorities, the dispossession of the peasantry is central to the strategy of the Tunisian capitalist state. Trapped in a cycle of indebtedness and reliant on international loans, the state seeks to stabilize its foreign reserves by promoting export-driven foreign currency inflows. This broader economic framework has reshaped agriculture—the sector represents around 10% of gross domestic product (GDP), more than the tourism sector (GIZ, 2019)—and land management in Tunisia more profoundly than any formal trade agreement.

The cycle of foreign currency dependence has facilitated the exploitation of "potentially profitable" collective lands, which have been granted or leased under long-term, favorable agreements to foreign capital, on the assumption that they were "empty" and "useless," echoing the colonial rhetoric of necessary and inevitable modernization. The development of the Deglet Nour industry exemplifies this trend, underscoring that Tunisia is far from an exception in this process. Marxist scholars argue that, in the 1980s, strategies of primitive accumulation resurfaced across much of the post-colonial world, playing a crucial role in the Cheapening of Nature. Federici (2019) refers to this shift as the "new enclosures," a process driven by the expansion of global capital, particularly in Africa, and facilitated by the intervention of International Financial Institutions (IFIs). Since the 2007–2008 commodity price spikes, this phenomenon has often been framed under the label of "land grabbing"—a term that frequently obscures its deeper historical dimensions (Edelman et al., 2013). In the case of southern Tunisia, as we have analyzed, it is essential to move past the immediate act of land acquisition and instead foreground the historical processes that underpin it. What is often presented as "new" is better understood as a "renewal" of longstanding patterns. As Edelman et al. (2013, p. 1521) point out, "the spaces in which land grabbing occurs have almost always been created and shaped by earlier processes of political contention, longstanding patterns of land tenure and use, and pre-existing social formations." Tunisia's colonial history, in particular, cannot be ignored, as the apparent novelty of contemporary land grabs and the push toward export-oriented agriculture can only be fully grasped through a historically grounded analysis.

From Common Land to a Fragmented Landscape: The Deglet Nour Export Market

According to French colonial literature, North African oases were depicted as "patches of greenery maintained by the relentless toil of men despite the dunes, the wind, and the sun ... herds led, as in biblical times, by nomads in search of meager pastures" (French minister of the Sahara, quoted by Gaudio, 1960, p. 104; Battesti, 2005, p. 12). In the European imaginary, the French Sahara evoked a religious landscape reminiscent of biblical descriptions and the supposed origins of Western civilization (Battesti, 2005). This colonial portrayal of an immaculate green haven—often framed as a miraculous anomaly—continues to shape contemporary international cooperation programs aimed at preserving oases, which are now framed as pristine ecosystems under threat from global warming and increasing drought in the region's semi-arid and arid climate (World Bank Group, 2019). As Battesti (2005, p. 12) observed, "the desert would not move, nor would the oases; everything would remain unchanged under the lethargy of an unforgiving sun." However, while today's oasis ecosystems in southern Tunisia are indeed endangered by worsening climatic conditions—forcing many former farmers to transition into wage labor—this environmental narrative alone does not fully account for their historical transformation. The shift from traditional agricultural practices to sites of value accumulation over the past 50 years also requires consideration. The question of what types of greenery oases have hosted in the past and host today will guide this section, politicizing crops and cultivation methods while avoiding the romanticization that, behind desert myths, has obscured attempts to appropriate and profit from fragile ecosystems in arid regions (Davis & Burke, 2011).

The expansion of date palm cultivation in southern Tunisia is closely linked to the introduction of a structured water distribution system in the region's oases. This invention is attributed to the Tunisian mathematician Ibn Chabbat, whose statue stands at the heart of Tozeur, a Saharan town in southwestern Tunisia, where he was born (1221–1282) (Omrani & Ouessar, 2008, p. 80). Among water management techniques, the traditional *foggara* irrigation system—a network of hand-dug tunnels that channel water from wells to agricultural fields, utilizing gravity to prevent evaporation in the desert heat—remains visible in some southern oases today. Local-scale methods of water management not only increased agricultural productivity but also contributed to "the maintenance of social equilibrium" (Omrani & Ouessar, 2008, p. 89). During this early period, a balanced production chain for date farming, collection, and trade emerged. However, as mentioned during a field interview by local actors advocating for oasis preservation—members of the Nakhla Association in Douz, located more than 350 km from the capital Tunis—the oasis, before its transformation into industrial monoculture, supported a far more diverse and ecologically balanced system of cultivation. This included a variety of fruits (e.g., peaches, plums, pomegranates, grapes, apricots, figs, mulberries, apples, pears, and even bananas) and vegetables (e.g., tomatoes, lettuce, garlic, onions, zucchini, squash, and chili) (Lassoued & Rejeb, 2016). The oasian soil also supported a wide variety of date palm species for centuries (Battesti, 2005). Starting from the 1970s, however, the industrialization of agricultural production significantly altered the local management of oasis ecosystems. The focus of date production shifted from meeting local needs to serving an external market increasingly interested in exotic, high-value fruits like dates. This transition led to the specialization of date cultivation for export, with a particular focus on the most profitable and labor-efficient variety: Deglet Nour.

Deglet Nour dates account for more than two-thirds of Tunisia's total date production and have become the country's third most significant food export, following seafood products and olive oil (Bachta et al., 2006). The sector expanded from 46,000 tons in 1980 to 113,000 tons in 2001 (Sai & Rhouma, 2005; Scanagri, 2004; Bachta et al., 2006) and further increased to 360,000 tons in 2023 (Agence Ecofin, 2023). As of October 2023, Deglet Nour dates constituted approximately 80% of Tunisia's total date exports (ONAGRI, 2023). Production is concentrated in two main governorates: Kebili, which accounts for 60% of total output, and Tozeur, which contributes 30% (GIZ, 2018). Tunisia consistently ranks among the world's top five date-exporting countries, alongside Saudi Arabia, Israel, Iran, and the United Arab Emirates (World Bank Group, 2025). The majority of Tunisian date exports are destined for the European market (approximately 50% of total exports) and Morocco (around 20%) (ONAGRI, 2023). Notably, this growth has primarily resulted from the expansion of cultivated land—particularly through so-called "extensions" surrounding the traditional oasis core—rather than improvements in productivity (Bachta et al., 2006). What was once a mosaic of diverse ecosystems defining the southern Tunisian landscape has been transformed into a monoculture system, altering the region's socio-ecological relations. This transition has led to intensified water resource exploitation, resulting in increased salinization due to the progressive depletion of aquifers (Hachicha & Ben Aissa, 2014). Additionally, shifts in land tenure have enabled land appropriation, turning collective farmers into wage laborers under the control of intermediaries in the export sector.

The Kebili region in southern Tunisia has experienced significant shifts in its oasis ownership structure, reflecting broader trends of privatization and market liberalization. Mekki et al. (2022) observe that by the late 1970s, various factors contributed to a shift toward privately managed oasis extensions outside traditional boundaries. In Kebili, rural areas are collectively managed through a complex system of tribal and community representation via councils at local, district, and governorate levels. Whereas these councils assign land allocation titles, they do not grant individual ownership. These titles, which can be transferred through sale or inheritance, enable entrepreneurs to access land either directly from the councils or by consolidating adjacent plots. Initially, private extensions outside traditional oases were not recognized by the administration.

Although formally classified as collective property, many uncultivated lands on the periphery of historical oases were gradually appropriated and developed by farming families, eventually becoming de facto hereditary holdings treated as private assets. The socio-spatial fragmentation of the oasian ecosystem is a direct consequence of the decline in communal land management. As land increasingly shifted to private, family-controlled management, often passed down through inheritance, it became fragmented into smaller, less productive plots. This fragmentation, typically resulting in plots no larger than one or two hectares, led to the abandonment of many parcels and a decline in agricultural productivity, as confirmed by local farmers. This phenomenon, commonly referred to as *parcellisation*, represents the division of family-run land into economically unviable units. As large-scale production becomes dominant, smaller producers face significant challenges in maintaining profitability, often prompting them to seek work outside the agricultural sector. These areas, located outside traditional oases and converted for palm tree production, account for 85% of the region's date output and 80.7% of Tunisia's total production, making them the largest illicit agricultural areas (in administrative terms due to the absence of formal land titles and regulatory approval, placing their use outside the official legal framework) dependent on groundwater (Mekki et al., 2022).

The rapid growth of Deglet Nour production, coupled with increasing land fragmentation—which primarily benefits export producers—has led to significant challenges, particularly regarding groundwater overuse and reliance on a single crop variety. The evolution of water management is intrinsically linked to land use practices. While traditional and government-established oases are collectively managed by water user associations, illicit drilling outside these areas has exacerbated groundwater depletion, threatening not only date production but also the survival of the oasian ecosystem (Mekki et al., 2022). The once diverse oasis ecosystem, sustained by natural springs, has been replaced by monoculture agriculture reliant on illicit wells. As Hachicha and Ben Aissa (2014, p. 777) observed, "since 1988, all natural springs were depleted, except for some mountain oases (Tamerza), and were replaced by drilled wells." A critical consequence of this shift is that "with the decrease of the water level of the deep aquifer, the water of the shallow groundwater can infiltrate and contributes to the salinization of the lower aquifers" (Hachicha & Ben Aissa, 2014, p. 778). This transformation marks the temporal establishment of a new commodity frontier in the capitalist Mediterranean, driven by the commodification of the oasis's sole product: the Deglet Nour date. Traditional farming practices, which evoke the customary infrastructure still present in the "mother" oases, have been replaced by the scalable mechanisms of global food value chains, regulating a horizontal system built on relationships and knowledge that is determined by price (Le Gal et al., 2008; and Bachta et al., 2006). Export prices have reversed the hierarchical relationship, leaving the organization of the market in the hands of exporters rather than producers, who bear the cost of declining product quality. A population once composed of local farmers and nomadic breeders, alongside the complex crop ecosystem of traditional oases, has been immobilized and put to work. Collectors, intermediaries, and exporters are now the key actors in the palm groves, where cheap labor is scarce. The shift to date monocultures has fundamentally altered the relationship between local communities and their environment, which, according to local actors interviewed during fieldwork, has become inhospitable and unsustainable. As a representative of the Nakhla Association stated during an interview conducted in Douz in November 2022, "There is a big difference between desertification and the encroachment of sand. Desertification is soil degradation, the loss of water resources, the degradation of plant cover, and even a reduction in population density. Stopping the sand is not enough to protect this fragile environment." Indeed, the expansion of palm trees is often justified as a means to combat desertification, with the advancement of sand halted by the cultivation of palm trees, as confirmed by local farmers. However, the policies promoting the greening of the desert seem to conceal a profit-driven agenda rather than serving purely environmental goals.

From Mobility to Immobility: Southern Tunisians in the Climate Migration Discourse

The transformation of socio-ecological relations in Southern Tunisia has been profoundly shaped by the distribution of profits generated by the export-import sector. This economic restructuring has contributed to what the Tunisian scholar Sadri Khiari describes as an "internal frontier" dividing the post-independence nation. This frontier separates the North-Eastern coastal region—where a new urban, economic, and political elite has emerged—from the South-Western inland areas, where rural communities have progressively lost access to land and economic mobility, ultimately threatening their livelihoods (Khiari, 2003, pp. 98–100). This duality opposes "*the export-oriented economic sphere, which benefits from a privileged allocation of resources, to the inward-looking sphere, destined to be marginalized by the pressures of international competition*" (Khiari, 2003, p. 98). One of the most tangible consequences of this socio-geographical rupture has been large-scale internal migration from inland rural areas to the cities on the coast, a process in which Southern Tunisia has played a significant role. As reported by the last general census of population and housing (INS, 2014), the trend in Tunisia, as elsewhere, is toward urbanization: its rate has risen steadily over the years, from 40% in 1966 to 52.8% in 1984, reaching 67.7% in 2014. This rural exodus intensified from the 1960s (Auclair, 2001), prompting demographic studies to address a question posed by Mahmoud Seklani at the start of that decade: "*How has the desert always been the provider of men for the cités (suburban buildings)?*" (Seklani, 1960, p. 501). Seklani's explanation adopts a depoliticized lens on migration, framing it as the inevitable outcome of demographic growth. Rooted in colonial epistemologies, this perspective abstracts migration from its broader socio-ecological and political context, thereby obscuring the historical processes of marginalization and dispossession that have shaped southern Tunisia (Seklani, 1960, p. 501).

Since the onset of the rural exodus, internal migration within Southern Tunisia has increasingly translated into external migration. The loss of land and living spaces, driven by the declining habitability of these regions, remains a pressing issue for Tunisia's rural and agricultural populations, particularly for younger generations. Emigration—whether to urban centers or foreign countries—has become a tangible and often necessary life prospect (Ramos, 2019). Data indicate that the South-Western region, including the governorate of Kebili, is not experiencing overpopulation but rather depopulation and a migration deficit, according to a study commissioned by the International Organization for Migration (IOM) (Bounouh & Gsir, 2017). However, the report identifies "unemployment" and "underemployment" as the "two main push factors for migration" (Bounouh & Gsir, 2017), without addressing the historical process of dispossession of the only means of production for a mobile rural and pastoralist population—land—and the generational consequences of proletarianization. Interviews conducted during fieldwork in the southwest of the country, highlight that mobility, whether through informal trade with neighboring countries (Algeria and Libya) or external migration, is perceived as the only sustainable response to the degradation of a territory's habitability. In Kebili, the interviews underscored how families lack male children under the age of 30, who, when present, have nonetheless tried to leave multiple times, only to encounter border policies imposed on Tunisia by the European Union and enforced by local officials. This context of constrained mobility intersects with shifting labor dynamics: the feminization of the agricultural workforce is not incidental but reflects broader patterns of precarity and social reproduction. Women constitute 50.4% of the rural population and represent approximately 35% to 40% of agricultural workers, underscoring their central role in sustaining both plantation production and household economies (Gharbi & Aribi, 2025). Those who manage to escape the desertification process often continue to cling to mobility: "Without Algeria nearby, we could not work," explained several informants from the Southwest-Algeria border area, who survive thanks to informal cross-border trade. The ongoing desertification of the oasis ecosystem and its reduction to a handful of businesses that attract export sectors, such as the example of Tunisian dates, is not merely an environmental phenomenon but a broader crisis of habitability rooted in the degradation of the natural environment and the proletarianization of rural inhabitants. This habitability crisis is often reduced either to a purely climatic issue—focused on rising temperatures and their consequences—rather than addressing the economic-capitalist orientation of the post-independence state, or to a purely economic pull-factor of migration. The example of the South-West of Tunisia illustrates that the migration phenomenon can no longer be understood purely from an anthropocentric perspective that rigidly separates natural and human dynamics. Instead, it must be reinterpreted also as a historical process within the context of commodity frontiers delineation, which inherently entwines social, ecological, and economic relations.

Conclusions

Since the colonial period, Tunisian territory—particularly its agricultural regions—has been shaped to serve agricultural value chains governed by the metropolis-colony power dynamic. Within this framework, profits are accumulated by the companies managing these chains, while the socio-ecological costs are borne by the colonial territory or, in the post-colonial context, the periphery. This power relation is neither an inevitable fate nor a natural given; rather, it is the product of the organization of the capitalist ecology-world, which is driven by the imperatives of wealth accumulation through the systematic devaluation of human and non-human natures, including the exploitation of living labor. The perpetuation of this dynamic of cheapness necessitates the continuous reconfiguration of both the colonial balance of power and the socio-ecological conditions characteristic of primitive accumulation. This includes, to varying degrees of coercion, the displacement of populations as their living environments become increasingly hostile or uninhabitable. Within this overarching long-term structure, agricultural production functions as a mechanism of capitalization that undermines local populations and their ecosystems. Contemporary climate change processes are exacerbating this trend, further entrenching the cheapening of agricultural lands and labor. Additionally, Mediterranean migration policies contribute to this process by further weakening Tunisia's rural communities, reinforcing their positioning as a sacrificed population within the broader capitalist world-ecology.

References

Abou, S. (2021, August 11). Initiatives européennes de l'UE en Tunisie. *Bilateral.org*. https://www.bilaterals.org/?tunisie-ue-initiatives-europeennes

Agence Ecofin. (2023, October 5). Tunisie: Hausse de 6 % de la production de dattes à 360 000 tonnes en 2023. *Agence Ecofin*. https://www.agenceecofin.com/fruits/0510-112414-tunisie-hausse-de-6-de-la-production-de-dattes-a-360-000-tonnes-en-2023

Auclair, L. (2001). Population et désertification en Tunisie au cours du XXe siècle. In L. Auclair, P. Gubry, M. Picouët, & F. Sandron (Eds.), *Régulations démographiques et environnement: actes des 6èmes journées démographiques de l'ORSTOM* (pp. 237–245). IRD/CEPED. https://horizon.documentation.ird.fr/exl-doc/pleins_textes/pleins_textes_7/carton07/010024401.pdf

Auzary-Schmaltz, N. (2008). Le régime foncier tunisien: Origines et évolution après l'indépendance. In C. Eberhard (Ed.), *Law, land use and the environment: Afro-Indian dialogues*. Institut Français de Pondichéry. https://doi.org/10.4000/books.ifp.3916

Ayeb, H., & Bush, R. (2019). *Food security and revolution in the Middle East and North Africa: Agrarian Questions in Egypt and Tunisia*. London: Anthem Press.

Bachta, M., Legal, P.-Y., Rhouma, A., & Kuper, M. (2006). From water to dates: An overview of the Tunisian date industry and prospects for interventions. In *Water Economies in Irrigated Systems in the Maghreb: Second Regional Workshop of the Sirma Project, Marrakech, Morocco, May 29–31.*

Battesti, V. (2005). *Jardins au désert: Évolution des pratiques et savoirs oasiens, Jérid tunisien.* IRD Éditions. https://horizon.documentation.ird.fr/exl-doc/pleins_textes/divers11-10/010035134.pdf

Ben Saad, A., Abaab, A., Bourbouze, A., Elloumi, M., Jouve, A.-M., & Sghaier, M. (2010). La privatisation des terres collectives dans les régions arides tunisiennes: Contraintes socio-économiques et impact sur l'environnement. Cas de la région de Tataouine, sud tunisien. Comité technique « Foncier et Développement ».

Besky, S. (2022). Reproducing the plantation. *Annals of the American Association of Geographers, 114*(10), 2212–2215. https://doi.org/10.1080/24694452.2022.2147050

Bounouh, A., & Gsir, S. (2017). Migrations et environnement en Tunisie: Relations complexes et défis pour le développement. Study conducted under the joint project of IOM/PNUD, "Intégrer la migration dans les stratégies nationales de développement," funded by the Swiss Agency for Development and Cooperation (SDC). https://tunisia.iom.int/sites/g/files/tmzbdl1056/files/documents/Etude%20migration%20environnement%20%28A4%29%20-%20FINAL.pdf

Bousnina, A. (2024). The migration of Tunisians: Determinants and effects. *Revue Africaine Des Sciences Humaines Et Sociales, 5*, 42–66. https://doi.org/10.34874/PRSM/rashs-n5.45856

Davis, D. K., & Burke, E., III (Eds.). (2011). *Environmental imaginaries of the Middle East and North Africa*. Ohio University Press.

Davis, J., Moulton, A., Van Sant, L., & William, B. (2019). Anthropocene, capitalocene, ... plantationocene? A manifesto for ecological justice in an age of global crises. *Geography Compass, 13*(5), e12438. https://doi.org/10.1111/gec3.12438

Edelman, M., Oya, C., & Borras, S. M., Jr. (2013). Global land grabs: Historical processes, theoretical and methodological implications and current trajectories. *Third World Quarterly, 34*(9), 1517–1531. https://doi.org/10.1080/01436597.2013.850190

Federici, S. (2019). *Re-enchanting the world: Feminism and the politics of the commons.* PM Press.

Gachet, J.-P. (1987). "L'agriculture : discours et stratégies." In Tunisie au présent: une modernité au-dessus de tout soupçon ? Nouvelle édition, edited by Institut de Recherches et d'Études sur le Monde Arabe et Musulman. Aix-en-Provence: Institut de Recherches et d'Études sur le Monde Arabe et Musulman. ISBN 9782271081278.

Gaudio, A. (1960). *Le Sahara des Africains.* Paris: Éditions René Julliard, Collection d'Histoire et Voyages.

Gharbi, I., & Aribi, F. (2025). Accès des femmes à la terre agricole en Tunisie. Presented at the Third Arab Land Conference, Rabat, Morocco, February 10–20. https://arablandinitiative.gltn.net/sites/default/files/women-s-access-to-agricultural-land-in-tunisia_aribi.pdf

Giudice, C. (2009). Législation foncière et colonisation de la Tunisie. In S. El Mechat (Ed.), *Les administrations coloniales, XIXe-XXe siècles* (pp. [pages if known]). Presses universitaires de Rennes. https://doi.org/10.4000/books.pur.104456

GIZ. (2018, June). *Tunisie: Amélioration du système d'approvisionnement d'un des leaders mondiaux en export de dattes.* https://www.giz.de/en/downloads/GIZ_SVAAA_Factsheet-Tunesia-Dates_FR.pdf

GIZ (Deutsche Gesellschaft für Internationale Zusammenarbeit). (2019, November). *Tunisian agri-food trade: Policy brief.* Bonn and Eschborn. https://www.giz.de/de/downloads/GIZ_SVAAA_Policy-Brief_Tunesia-AgTrade_EN.pdf

Hachicha, M., & Ben Aissa, I. (2014). Managing salinity in Tunisian oases. *Journal of Life Sciences, 8*(9), 775–782. https://doi.org/10.17265/1934-7391/2014.09.007

Holm, H. M., & Santmyer, C. (1964). *Agriculture in Tunisia: Organization, production, and trade* (Miscellaneous Publications 316358). United States Department of Agriculture, Economic Research Service.

Institut National de la Statistique (INS). (2014). *Recensement Général de la Population et de l'Habitat 2014, Volume 5: Caractéristiques démographiques et économiques.* Tunis. https://www.ins.tn/sites/default/files/publication/pdf/RGPH%202014-V5.pdf

Khiari, S. (2003). *Tunisie: Le délitemnt de la cité - coercition, consentement, résistance.* Éditions Karthala.

Lassoued, K., & Rejeb, H. (2016). The oasis of Gabes in Tunisia: An endangered authenticity. *The International Journal of Social Sciences and Management, 6* (7), 80–87.

Le Gal, P.-Y., Gendre, L., & Rhouma, A. (2008). Impacts de la chaîne d'approvisionnement export sur la valorisation de l'eau par les dattes dans les oasis du sud-tunisien. In M. Kuper & A. Zaïri (Eds.), *Economies d'eau en systèmes irrigués au Maghreb: Actes du troisième atelier régional du projet Sirma, Nabeul, Tunisie, June 4–7 (pp. [page numbers if known]).* Cirad.

Lebdi, F., & Maki, A. (2023). *La sécheresse au Maghreb: Diagnostic, impacts et perspectives pour le renforcement de la résilience du secteur agricole.* FAO. https://openknowledge.fao.org/server/api/core/bitstreams/3099c078-a57d-47b9-a549-8fede794e08e/content

Mahjoub, A. (1987). Économie et société: La formation du « sous-développement ». L'évolution socio-économique de la Tunisie précoloniale et coloniale. In M. Camau (Ed.), *Tunisie au présent.* Institut de recherches et d'études sur les mondes arabes et musulmans. https://doi.org/10.4000/books.iremam.2551

Mekki, I., Ferchichi, I., Taouajouti, N., Faysse, N., & Zairi, A. (2022). Oasis extension trajectories in Kebili territory, Southern Tunisia: Drivers of development and actors' discourse. *New Medit, 21.* https://doi.org/10.30682/nm2205f

Midnight Notes Collective. (1990). Introduction to *The New Enclosures. Midnight Notes, 10*, 1–10.

Moore, J. W. (Ed.). (2016). *Anthropocene or Capitalocene? Nature, history, and the crisis of capitalism*. PM Press.

Moore, J. W. (2021). Climate, class & the great frontier: From primitive accumulation to the great implosion. *World-Ecology Research Group, Binghamton University*. https://jasonwmoore.com/wp-content/uploads/2021/09/Moore-Class-Climate-and-the-Great-Frontier-circulating-September-2021.pdf

Omrani, N., & Ouessar, M. (2008). Historical and contemporary perspectives of water culture in Tunisia. In M. El Moujabber, M. Shatanawi, G. Trisorio-Liuzzi, M. Ouessar, P. Laureano, & R. Rodríguez (Eds.), *Water culture and water conflict in the Mediterranean area* (pp. 75–84). CIHEAM. *Options Méditerranéennes: Série A. Séminaires Méditerranéens*, No. 83.

ONAGRI. (2023, October). Flash sur les exportations des dattes tunisiennes. http://www.onagri.nat.tn/uploads/filieres/dattes/Flash_Dattes_nov-2023_Fr.docx.pdf

Ramos, R. (2019). Migration aspirations among youth in the Middle East and North Africa region. *Journal of Geographical Systems, 21*, 487–507. https://doi.org/10.1007/s10109-019-00306-1

Sai, M. E., & Rhouma, A. (2005). Le secteur des dattes en Tunisie : les contraintes et les opportunités dans le nouveau contexte économique. In F. Cena, R. Gallardo, M. Elloumi, & M. E. Sai (Eds.), *Les défis de la terre: l'agriculture en Espagne et en Tunisie face aux défis de la libéralisation* (pp. 223–240). Tunis: Cérès – IRESA.

Scanagri. (2004). Étude des marchés des produits du secteur du palmier dattier au Maghreb. Rapport final. Projet FEM-PNUD RAB/98/G31 « Gestion participative des ressources génétiques du palmier dattier dans les oasis du Maghreb ». Tunis: IPGRI, PNUD, FEM/GEF, INRA Tunisie, INRA Algérie, INRA Maroc.

Seklani, M. (1960). Villes et campagnes en Tunisie: Évaluations et prévisions. *Population, 15*(3), 485–512. https://www.persee.fr/doc/pop_0032-4663_1960_num_15_3_6657#pop_0032-4663_1960_num_15_3_T1_0498_0000

Tsing, A. L. (2013). Sorting out commodities: How capitalist value is made through gifts. *HAU: Journal of Ethnographic Theory, 3*(1), 21–43. https://doi.org/10.14318/hau3.1.003

Tsing, A. L. (2017). *Le champignon de la fin du monde: Sur la possibilité de vivre dans les ruines du capitalisme*. Éditions La Découverte.

World Bank Group. (2019, May 24). *Renewal of Tunisian oases creates jobs; boosts climate resilience*. https://www.worldbank.org/en/news/feature/2019/05/24/renewal-of-tunisian-oases-creates-jobs-boosts-climate-resilience

World Bank Group. (2025). *Exports of dates in 2023*. World Integrated Trade Solution. Retrieved February 24. https://wits.worldbank.org/trade/comtrade/en/country/ALL/year/2023/tradeflow/Exports/partner/WLD/product/080410#:~:text=In%202023%2C%20Top%20exporters%20of,88%2C153.17K%20%2C%2018%2C436%2C300%20Kg

Extractivism as Predatory Accumulation. On Water, Labor, and Accumulation in Strawberry Production in Southern Spain

Susana Narotzky and Natalia Buier

Introduction

The province of Huelva, in southern Spain, is one of the major world exporters of strawberries and other red fruits. Local agriculture has recently been on the media for two contentious issues, on the one hand, the hyper-exploitation of migrant labor, and, on the other hand, the over-extraction

S. Narotzky (✉)
University of Barcelona, Barcelona, Spain
e-mail: narotzky@ub.edu

N. Buier
Complutense University of Madrid, Madrid, Spain
e-mail: nbuier@ucm.es

S. Narotzky et al. (eds.), *Agricultural Extractivism in the Mediterranean Region*,
https://doi.org/10.1007/978-3-032-05599-6_5

and depletion of aquifers for intensive irrigation.[1] Delving into the socio-ecological order of intensive berry production in Huelva through the lens of agricultural extractivism, this chapter sheds light on the specificities of contemporary accumulation processes in agro-industrial production. The case of Huelva also allows us to interrogate the merits and limits of existing conceptualizations of agrarian extractivism that highlight its predatory nature.

We do so by addressing the contradictions that connect the accumulation of value, the reproduction of human and non-human nature, and the continuity of the political economic system. We have focused on water and labor which are crucial "resources" for the intensive production of berries as an export commodity. A historical and ethnographic perspective enables us to capture the changes in the socio-political and economic uses and values that are attached to these biophysical resources in the process of producing them as such. In the first section, we discuss the usefulness of concepts referring to the notion of predation through which the distinctiveness of agrarian extractivism might be captured. The second and third sections introduce the case study by focusing on the production of two main agricultural inputs, labor and water. In search of explanatory arguments, the fourth section analyzes it against the various theoretical frameworks of "productive destruction." In the final section, we conclude that a crucial aspect of value extraction and accumulation is the (dis)connection between production and reproduction, namely the fact that reproduction of the materials or energy used in production go often unaccounted for in market exchange.

The chapter's methodology combines different research methods. On the one hand, extensive documentary research including statistics from various sources, institutional reports, grey literature, and printed and visual media material, and, on the other hand, ethnographic fieldwork in a strawberry and berries production region bordering the Doñana wetlands in Andalusia. The documentary research refers to the 2019–2023 period, while the ethnographic research extends for 10 months from the fall of 2021 to the summer of 2022.

[1] https://www.theguardian.com/global-development/2023/mar/31/abusive-working-conditions-endemic-in-spains-strawberry-farms-report-claims, https://www.lemonde.fr/en/environment/article/2023/04/28/intensive-strawberry-farming-is-threatening-spains-donana-wetlands-a-world-heritage-site_6024728_114.html

Natalia Buier conducted research in the intensive berries production area of Huelva bordering the protected wetlands of Doñana. An additional two months were spent doing archival research and interviews in the capital town of Andalucía, Sevilla. While residing in the two municipalities, she carried out participant observation in everyday settings and conducted interviews with a wide range of actors involved in and affected by intensive agriculture groundwater exploitation (farmers, laborers, union organizers, water managers, administrators and policy experts, environmental activists, etc.). This long-term intensive ethnographic fieldwork has been complemented by the continuous gathering and analysis of documentary and archival information.

Documentary research has been carried out by Susana Narotzky starting in 2019 and up to the present. An archive of over 500 items including legal documents, official reports, grey literature, and media has been compiled, categorized, and analyzed. The extended use by social agents of social networks and video clips in news media has proven to be a useful tool to access secondhand interview material, especially during the pandemic. Both authors have collaborated intellectually during the period of research and analysis and this article is the result of this collaboration.

Agricultural Extractivism and Value Production and Destruction

The concept of agricultural extractivism, as presently understood, emerged from the theorization of "new extractivisms" following the commodity boom at the beginning of the twenty-first century. It is connected to the efforts of Latin American scholars to understand the characteristics of accumulation models supported by the so-called "pink-tide" governments, in which natural resource extraction, primary commodity exports, state appropriation of resource rents, and the state's objective to reduce poverty played a central role (Svampa, 2019; Gudynas, 2010; Acosta, 2013). The study of new extractivism expanded the field beyond its traditional scope of hydrocarbon and mineral extraction to biofuel plantations and flex crops exports. It also pointed at the differences between previous extractive processes and new extractivisms. Unlike colonial and neocolonial extraction of raw materials from "peripheries" that resulted in accumulation of wealth at the "center," neo-extractivism presented redistribution as the aim of resource rents accumulated by the state through extractive endeavors and primary commodity exports. New extractivism articulated a strategy of "national economic development" through primary commodity exports with a focus on reducing poverty through redistributive policies (Petras & Veltmeyer, 2014). Attaching itself to redistributive agendas, neo-extractivism appears as a form of resource nationalism regarded as a distinguishing feature of the early twenty-first-century reprimarization of Latin American economies (Svampa, 2019). Yet, it has also been defined as "authoritarian corporative populism" in that state actors, elites, and agroextractivist capital combine the manufacturing of popular consent with selective repression and the use of violence to expand commodity frontiers (Alonso-Fradejas, 2024).

Agricultural extractivism acquires its distinguishing features within the framework of the new-extractivism. Thus, initially agricultural extractivism is tied to plantation crops, such as sugarcane, soy, palm oil, that are defined as "flex" crops because of their various uses (food, fuel, feed, industrial), which can fend off price volatility in commodity markets and profit from peak prices. These crops have been described as linked to food security, energy transition, and climate change mitigation in various ways, and also as providing state revenues that could be redistributed to further social development (Borras Jr et al., 2014). Recent approaches to agrarian extractivism have proposed an expanded notion that stresses the fact that "extractivism represents *production without reproduction*" (Ye et al., 2019, p. 3, emphasis in original). Without forgoing the large-scale, intensive monocrop production for export meaning, current theorization has underlined monopolistic control over flows, environment destruction, resource depletion, and degradation of work in "an organized, and internally coherent, system for ongoing value extraction" (Ye et al., 2019, p. 5). Indeed, one of agricultural extractivism's main characteristics is the tendency to destroy its own resource base through non-reproduction. Yet, agricultural extractivism's resilience seems to be just as much about its ability to meet depletion with an incessant process of converting human and non-human nature into resources. This is not simply about identifying new labor and water reserves, but involves a multiscalar process of transforming these—materially and ideologically—into agricultural inputs. Locked into a constant search for new deposits of labor and water, agricultural capital must redraw entire geographies of reproduction in order to channel them into the valorization process. We consider the case of Huelva helpful for further developing an expanded notion of agricultural extractivism because of its material and ideological expelling of reproduction from the valorization process (and its accompanying contradictions), even as labor and water are produced as agricultural inputs. Both of these processes push us to reflect on the relation between predation and accumulation, or what we describe as the interlocking aspects of value production and destruction.

Agricultural extractivism, as an accumulation model accompanying corporate-led, external-input plantation agriculture (McKay et al., 2021), has been described as a "predatory, life-purging" system of social relations (Alonso-Fradejas, 2021). The metaphor of predation is a powerful one that points us to the relationship between production and destruction of value, a defining aspect of agricultural extractivism. Hence, we wish to explore possible connections between predation and concepts that share a semantic space, in particular those of creative destruction, productive consumption, accumulation by dispossession, and predatory accumulation. To think of agricultural extractivism as predatory underlines its tendency to destroy ecological and social reproduction, yet we wish to investigate the various declensions that this destructive aspect may hold to better capture the specificity of the Huelva case study that we present.

Destruction and its Declensions

Perhaps the concept most widely used and misused is Schumpeter's notion of "creative destruction" (Schumpeter, 1994). For the purposes of this chapter, we want to highlight one overlooked aspect of the processes of innovation characteristic of capitalism's evolution that Schumpeter captures with the notion of creative destruction, namely their temporal logic. For Schumpeter, creativity, in the form of innovation, implies a profound transformation of existing practices, one that eventually destroys the old ones. The creative upheaval fosters competitive advantage that generates the destruction of obsolete firms. Both technological and organizational changes can be innovative and destructive of previous processes. For agriculture in Huelva, technological innovations such as fertilizers, pesticides, irrigation and greenhouse infrastructures, digitalized control, plant varieties, biological pest control, and organizational ones such as migrant program labor recruitment and commercializing hubs, all have produced competitive advantage at different times and posed a threat to less "innovative" producers. As we will see below they have also proven destructive in terms of environmental and labor reproduction which connects this Schumpeterian idea to local agrarian extractivism.

Classic "productive consumption" reverses the previous temporal framework as production is predicated on the consumption—that is, on destruction through incorporation—of land, labor, raw materials, and tools into the commodities during the capitalist process of production. We will refer to this as "productive destruction" when stressing its predatory aspect. Marx (1982, pp. 342, 367, 416) likens the capitalist to a vampire, sucking the blood of laborers and hence producing death, but capital also thrives on the combination of dead and living labor where "living labour appears merely as a means to realize objectified dead labour [in the means of production], to penetrate it with an animated soul while losing its own soul to it" (Marx, 1993, p. 461; Neocleous, 2003). The objective of this productive destruction is the growth of a particular form of material wealth—capital—that enables those who own and control it to syphon the lives of those who do not into channels that expand that wealth. Here, labor consumption and, specifically, its precarious reproduction is the aspect that we wish to connect with the case of Huelva's agricultural extractivism. The concept of "productive consumption" acquires its full meaning as a part of the broader process of capital valorization. Productive consumption, in Marx's reading, should be understood as a moment in the total circulation of capital. And here the notions of over-accumulation and accumulation by dispossession come into play.

Crises of over-accumulation of capital, that is, periods when capital cannot be productively invested, result in different forms of devaluation (of labor power, of money, of commodities, of fixed capital) that destroy excess capital. Crises can often be understood as processes of destruction of value that prepare the ground for a further cycle of valorization. This violent destruction is not external to the reproduction of capitalism but a condition of its continuity (Harvey, 1999, p. 202). Harvey (2003) underscores another aspect of destruction, one which is directly predatory and linked to the accumulation of value that he defines as "accumulation by dispossession." Marx's rendering of an initial "primitive" stage of predatory violence in the origins of capital accumulation (considered from an evolutionary viewpoint), and Luxemburg's widening of the process to colonial expansion, described this predatory process as one extracting value from "outside" the capitalist system. Harvey, instead, extends the notion to forms of accumulation that are produced—he uses the term "manufactured"—"inside" capitalism as an aspect of its reproduction (Harvey, 2003). By pointing at the many forms in which present-day capitalism uses violent extraction such as forceful privatization, financial entrapment, depletion of natural resources, and people's devaluation as a regular mode of accumulation, Harvey also underscores the *internal* process of producing kinds of human and non-human nature that may be categorized as preys: "Capitalism internalizes cannibalistic as well as predatory and fraudulent practices" (Harvey, 2003, p. 148). In the Huelva case study below, the devaluation of land—through the depletion of aquifers—and labor, through regulatory and ideological means, creates the conditions for further cycles of accumulation as it extracts resources whose reproduction is not accounted for.

Finally, a word on Philippe Bourgois' concept of "predatory accumulation" that he also describes as "lumpen accumulation" (Bourgois, 2018, p. 390). The gist of the argument is that the illegal narcotics economy accumulates value through using labor expelled from the licit circuits of the economy—that is, lumpenized populations—and thrives by killing or maiming both consumers and labor. The stress on "illegality" and on death underscores the aspect of producing internal devalued others that may feed accumulation as well as absorb idle capital (à la Harvey). Yet, the deadly aspect is important and reflects other accounts of predatory and destructive processes at the base of capital accumulation.

Indeed, a connected debate has developed around the concept of "necroeconomics" that focusses on the predatory angle of death and is inspired by Achille Mbembe's "necropolitics." Mbembe depicts colonial occupation processes where the colonized were relegated "to a third zone between subjecthood and objecthood" (Mbembe, 2003, p. 26, see also Banerjee, 2008). Speaking of contemporary spaces of war and resource extraction in post-colonial situations, he describes extreme technologies of destruction where "massacre" and "survival" are the expressions of a new form of governmentality *from which profits derive*—mining, weapons, etc. (Mbembe, 2003, pp. 33–35). Underscoring the continued relevance of "making die" in connection to processes of extraction, exploitation, and dispossession brings to the fore how death is not only a consequence but a *necessary condition* for capital accumulation (Montag, 2005; Murphy, 2017). This connects with the extremely precarious reproduction of agricultural labor in Huelva one of the central aspects of local agricultural extractivism.

We will now present the case of intensive red fruits cultivation, human exploitation, and water depletion in Huelva, following up with a discussion on what we find to be the more productive route to theorizing the interlocking aspects of value creation and destruction in the framework of an expanded notion of agrarian extractivism that stresses the non-reproductive aspect of the process.

The Productive Destruction of Labor

We have analyzed an area known as Corona Norte ("Northern Crown") of the Doñana Natural Space, a protected wetland. The Corona Norte comprises a string of municipalities (Almonte, Lucena del Puerto, Bonares, Rociana del Condado, Moguer) where the dominant economic activity is berry production. These municipalities form Europe's most important berry exporting area. In 2023, Spain was the second largest world exporter (in value) of strawberries (19.4%) after Mexico (31.5%), exporting mainly to Germany (26.5%), the United Kingdom (16.2%) and France (13.6%).[2]

The origins of berry cultivation in the province of Huelva are to be found in the 1960s, when experiments with strawberry culture proved successful among a host of agronomic trials sponsored by the Food and Agriculture Organization (FAO) and intended to transform an allegedly unproductive region into a competitive agricultural producer. Strawberry farming soon became established by a private entrepreneur in the Moguer municipality. In the 1970s, and especially during the 1980s, expansion of the crop occurred through the (un)regulated occupation of municipal land by unemployed people. The "right" to claim land for subsistence from the municipality was recognized at least since the eighteenth century (Ojeda Rivera, 2022 [1987]) and was granted in the form of "colonato," small parcels subject to a "canon" fee. Occupation of small parcels (average 3 ha) during the 1980s mostly in forested areas were a consequence of various factors that include (1) the displacement of previous "colonos" from their land to develop an industrial park in the early 1970s, (2) unemployment following the 1973 crisis, and (3) the success of the new intensive irrigated agriculture. Subsequently, the initial occupations and their access to water were legalized although encroachment upon nearby public domain remained a persistent feature (Márquez Domínguez, 1986, pp. 86–88; Jiménez Díaz, 1992; Reigada, 2022a, pp. 42–51).

Smaller farms initially relied on household labor, but increased productivity of the crop at the turn of the century made production more dependent on migrant work, initially that of day labor from other Andalusian

[2] OEC, https://oec.world/en/profile/hs/strawberries-fresh?cumulativeMarketShareSelected=share&yearSelector1=2021

provinces. As strawberry production expanded and tourism and construction opportunities grew, regional labor reserves became insufficient and in the 1990s the intensive phases of the production process became heavily dependent on male migrant laborers, both in regular and irregular situation. By the turn of the century, a wave of protests led by rural migrant workers, mostly Moroccans, revealed the growing costs of containment associated with using cheaper labor, that is, with producing and taking advantage of a devaluated labor pool. Simultaneously, the European Union was pushing for a tougher control of migration flows. In this context Spain's guest worker program, locally known as *contratación en origen* ("hiring at origin"), was born in 1999 setting national quotas, preference for countries with migration agreements, and strict residence and mobility limitations for labor (Márquez Domínguez et al., 2013; Gualda, 2008; Molinero Gerbeau, 2020).

For two decades, this program has provided Huelva's agribusiness with the bulk of the labor force required at harvest time, the most labor-intensive phase of berry production (Hellio & Moreno Nieto, 2021; Reigada, 2022b). Huelva became the main beneficiary of the hiring at origin program under the leadership of the municipality of Cartaya, whose socialist mayor, Juan Antonio Millán, actively promoted a multi-scalar coalition for its implementation. With EU funding from the AENEAS program,[3] the Cartaya administration initiated a pilot program for the import of labor from non-EU countries that consolidated circular migration. This also marked the decided feminization of berry picking. Between 2000 and 2004, the number of day laborers hired at origin multiplied by 35, primarily women of Polish and Romanian origin (Gualda Caballero & Ruiz García, 2004). After 2007, following Romania's and Poland's accession to the EU, contracting moved to Morocco. This is how Juan Antonio Millán, self-fashioned "inventor of hiring at origin," unabashedly recalls the process:

> In the year 2003, I made a consideration: circular migration is useful with people coming from third countries, because if a person is already a [European] Community member, he or she can move freely. Of course, in 2000 Polish women were third country, in 2000 and 2001, but in 2002 they were not, that is why we started to bring Romanian women, because

[3] https://eur-lex.europa.eu/LexUriServ/LexUriServ.do?uri=OJ:L:2004:080:0001:0005:EN:PDF

> Romania was a third country. And I thought, well, when Romania joins the European Union, where are we going to bring them from? From America? Flights are very expensive … but we already have Morocco, we are here, one hour away from Tangier, one hour by boat and three hours by bus, four hours…. So we have to set our sights on Morocco. (Juan Antonio Millán, personal interview, 2022)

However, by 2020, under the impact of the COVID-19 pandemic and political tensions between Spain and Morocco, the Huelva agricultural sector was faced with the threat of an insufficient and unreliable supply of cheap labor and thus began looking at the very places that had previously appeared too far. Since 2021, the hiring of Moroccan female workers has been supplemented with small quotas of workers from Ecuador and Honduras. Asked to comment on this most recent episode in redrawing the boundaries of the regional labor market, Juan Antonio Millán retorts:

> […] Morocco's government is unstable. In Morocco, moreover, there are more and more agricultural companies, many of them Spanish, and there are fewer and fewer women with the desire to come. That is why there is a need to look for new [labour] *deposits*.[4]

Much like in the case of water that we explore in the next section, the conditions for the exploitation of a "labour deposit" are not fixed, but determined by a variable constellation of political-economic factors and socio-technical tools that contribute to "produce" people as a labor "resource." Capital's free mobility and labor's contained mobility are both essential to dealing with changes in the social conditions governing the exploitability of labor deposits.

We have expressed elsewhere the arguments that employers use to explain their labor preferences in terms of ethnic, racial, cultural, and gender justifications that refer to docility ("not being confrontational"), resilience (supporting hard work), and productivity (skill, energy, speed) (Narotzky, 2021). This has increasingly favored the mining of a particular "deposit" of female labor characterized by such personal criteria as having children under 14 (Gualda, 2012), but also by biophysical and sociocultural qualities, as the following quote illustrates:

[4] "El gobierno de Marruecos es inestable. En Marruecos, además, cada vez hay más empresas agrícolas, muchas españolas, y cada vez hay menos mujeres con apetencia de venir y por eso hay que buscar nuevos *yacimientos*"

> Why [hiring] women and not men? Well, very simple. First, women have more endurance than men. A woman has a greater capacity for suffering than a man. A woman is more docile than a man. A woman is more selective than a man. More neat. (Francisco López, farmer, in Reigada, 2022a, p. 180)

Yet, preference is also voiced in terms of bonded labor:

> If right now all foreigners [workers] here in Palos had their papers [were regular residents], the campaign could end now, because now my women could say "Javier, there is little fruit here and I am going to go to Lleida to pick orange, or Murcia to pick tomato, or Almería." When they have their papers, you cannot stop them. What can you do? You have to pull up the strawberry plants. Otherwise, if they have a contract in origin [temporary labor program], they know that until the boss says the campaign is over, they cannot leave. (Javier Moreno, small farmer, in Reigada, 2022a, p. 174)

Put otherwise, labor must move, but not under conditions of its own choosing. "The key to circular migration and hiring at origin is that people return to their country" (Juan Antonio Millán, personal interview 2023), that is what makes them dependable and legally entitled to "repeat" a temporary contract the following season. Women who do not return are defined as "*fugadas*" (fugitive), a concept applied to escaped prisoners, they become "illegal" migrants. Legally prevented from changing employers and housed in containers on the farm, women workers hired at origin are subjected to a process through which their reproductive time is reduced to the necessary minimum.

Indeed, the reproduction of dependents is directly excluded in the migrant labor programs that bring Moroccan women for the strawberry harvest who have to rely on relatives to take care of their children back home, even as family care is the main incentive for their work and docility; everyday reproduction is indirectly excluded by pushing migrant workers to live out of sight, away from town centers in container housing or in self-constructed shanty settlements. Most migrant workers accept exploitation because they are responsible for their families abroad through remittances; at the same time, they have to maintain their energy and their value as labor otherwise they are "punished" for lack of productivity (Filigrana, 2021).[5] While capital strives to free itself from the costs of

[5] Workshop on precarious day labor in agriculture organized by Confederación General del Trabajo, Jornaleras de Huelva en Lucha, Fruita amb Justícia, CEMyRI Universidad de Almería, December 14, 2021

labor's reproduction, there is no divide between production and reproduction for migrant labor: their work is tied to the reproduction of life and their mobilizations all underscore the need for decent housing and the right to dignity.

The Productive Destruction of Water

The idea that the "regeneration" of the nation was predicated on making drylands productive through irrigation harked back to Enlightened intellectuals, such as Joaquín Costa (1846–1911) who proposed hydraulic policy as the salvation of the material and social problems of the country. As Ortí (1984) and others have underlined, this was an attempt to solve both the Agrarian Question (particularly salient after the 1880 agricultural crisis), and the Social Question (as the organizing of day laborers increased), in a way that would not imply the overturn of the land ownership structure. It is interesting to consider the problem of agricultural extractivism through "hydrosocial extractive territories" (Artiga-Purcell, 2022) because as a force of modernization irrigation has long been construed as a form of resource nationalism that would enable a redistributive program. The planning of irrigation at the national scale was taken up by governments of all persuasions and became a special material and symbolic asset of the Franco regime (Ortega Cantero, 1984; Barciela, 1996; Gómez Benito, 1994; Swyngedouw, 2015). Two issues seem to us important in this historical overview: the technosocial aspect of hydraulic policy that was clearly underscored early on as a "production of nature,"[6] and the co-dependence of irrigation, economic development, and social wellbeing. This, at the very least, raises some questions about the novelty of extractive redistributive agendas (e.g. new extractivism) and points to the possibility of revisiting some more recent phenomena through the old agrarian questions.

By the 1980s, strawberries had become a dominant crop in the Corona Norte area, and four decades later, strawberries and other berries constitute Huelva's major crop. Key to the consolidation of hyper-intensive agriculture in the region has been its heavy reliance on agrochemicals, irrigation, and cheap wage labor. Irrigation water in the Corona Norte area comes from the Doñana aquifer, also known as the Almonte-Marismas aquifer. This aquifer feeds the temporary and permanent ponds of the

[6] In Joaquín Costa's words: "Si en otros países basta con que el hombre ayude a la naturaleza. Aquí tiene que hacer más, tiene que crearla" (quoted in Ortí, 1984, p. 45)

wetland complex that support the reproduction of the park's wildlife. Extending over more than 3000 km^2, the aquifer overflows the boundaries of the protected area and can be extracted through relatively simple pumping well technologies (García Ruiz, 2018). The relentless expansion of berry cultivation in the last four decades is the main source of pressure on the aquifer, occasioning both quantitative and qualitative problems.

Two events led to an escalation of tensions surrounding groundwater extraction in the Doñana natural space. In 2021, the European Court of Justice (ECJ) declared Spain in breach of European water and environmental legislation, having failed to halt the deterioration of the Doñana park due to "excessive extractions of groundwater." In February 2022, despite the ECJ ruling, a regional bill, promoted by right-wing and far-right parties,[7] sought to expand the irrigated area in the proximity of Doñana by securing water rights for some 1500 hectares of land that were classified as illegally irrigated under the current legislation. Following the victory of the conservative Partido Popular (PP) in the autonomic elections of Andalucía, the bill obtained the backing of the regional government and became the centerpiece of a protracted negotiation between the regional and the left-of-center national government coalition.[8]

Opponents of the new bill, most prominently environmental organizations, considered it an assault on the arduously negotiated 2014 zoning plan known as "Plan de la Fresa"[9] which limits irrigation in the Corona Norte region through classifying land into irrigable, rain-fed, and forested areas. The Plan de la Fresa had legalized irrigated land productively in use before 2004 but lacking formal irrigation rights. Opponents to the 2022 new attempt to legalize unregulated irrigated lands argued this would be the final blow to an area already severely affected by water stress. In contrast, supporters of the new bill, mobilized by the agro-industrial lobby and right-wing political forces, argued that the farmers who were denied water rights in the 2014 plan were victims of a "historical injustice," by which they primarily disputed the limited retroactive character of the special zoning plan approved in 2014.[10] They additionally argued that water

[7] Partido Popular (PP), Ciudadanos and VOX

[8] Partido Socialista (PSOE) and Unidas Podemos

[9] Strawberry Plan

[10] The 2014 zoning plan for the "Corona Norte" area is subordinated to the 2004 zoning plan for the entire area of Doñana (POTAD, *Plan de Ordenación Territorial del Ámbito de Doñana*). The 2014 plan is referred as Strawberry plan (Plan de la Fresa) https://www.juntadeandalucia.es/organismos/fomentoarticulaciondelterritorioyvivienda/areas/ordenacion/actuaciones-supramunicipales/paginas/plan-corona-forestal-donana.html. Due to the

rights did not guarantee access to water, but only potential access provided sufficient resources were available.[11] Securing those resources, they argued, required the completion of several infrastructural projects that would enable the substitution of groundwater with surface water transfers. Thus, claims to infrastructural fixes are central to the agricultural lobby's arguments.[12]

In November 2023, negotiations between the regional and national governments resulted in a provisional agreement: the regional government committed to withdrawing the bill in exchange of a nationally funded investment plan destined for the berry producing areas near Doñana. The agreement between the national and regional authorities foresees financial incentives for farmers willing to convert irrigated land into forested areas. In exchange for halting irrigation and reforesting an area for a duration of 30 years—without losing their property titles—farmers in the Corona Norte are expected to receive a compensation of 100,000 euros per hectare (at a rate of 10,000 €/ha for a period of 10 years).[13] Hence, while the search for water scales up in infrastructural fixes, the fact of depletion is being met with a different range of strategies for revalorization. Agrarian capital confronts the cumulative depletive effects of its past activity through the speculative creation of fictitious water claims subsidized by the state if they renounce to them. Farmers will receive money provided they give up using water resources to which they are not entitled and/or which their intensive irrigation has depleted. As a result, their land

complexity of the area in the Corona Norte, the elaboration of this additional plan took 10 years. The land classification in the 2014 "Plan de la Fresa" retroactively granted water rights to all those whose lands were in cultivation up to the year 2004 (the year POTAD came into force) and excluded from legalization all lands put into cultivation afterwards. During the decade-long process of elaborating the "Strawberry Plan," illegal cultivation and irrigation continued to expand. By 2014 the difference between 2004 and 2014 was substantial, hundreds of newly irrigated hectares: these are the irrigators excluded from the Plan de la Fresa who contest it.

[11] The demand was for land to be classified as "irrigable" (*suelo regable*) that would be only actually irrigated in case of sufficient water resources, the actual allocation of water depending on the Guadalquivir Hydrographic Confederation.

[12] https://www.eldiario.es/andalucia/huelva/agua-andalucia-pide-portugal-presa-abastece-recorte-regadio-extremeno-castellano_1_10714284.html and

https://www.diariodehuelva.es/articulo/provincia/organizaciones-agricolas-lanzan-movilizacion-15n-recortes-agua-impuestos/20231107185530310674.html

[13] The agreement also establishes subsidies for the substitution of irrigated crops by rainfed ones (2500€/ha for a maximum period of 5 years) and conversion to ecological production (1000€/ha for a maximum period of 5 years).

will be revalued, whereas their intensive farming had previously devalued it. At the time of writing, it remains unclear how compensation for farmers without water entitlements will proceed, yet its application would mean, on average, a tenfold increase of the financial value of their legally un-irrigated land, as in Huelva dryland trades, on average, at prices 10 times lower than those of irrigable land, roughly 10,000 euros per hectare.[14] In this context, the agreement reached between the regional and national government must be understood as an attempt to revalorize assets at the point of their devaluation.

Conclusion: Productive Destruction between Extractive Predation and Accumulation

The case of berry production process in Huelva is similar to that in other regions of Europe (e.g. Italy, Portugal, France) and the world (e.g. California), albeit with its particular history. These processes have been analyzed innumerable times from the perspective of global commodity chains and accumulation by dispossession. Can our analysis provide a new angle by making use of the concepts of predation, productive destruction and extraction?

In our case study, we observe that different instances of what we may call predation are routinely present, for example, the occupation of public land by families claiming a right to a livelihood through farming, the use of groundwater from unregulated wells, and the employment of hyper-vulnerable labor (unregulated migrants, temporary migrant labor, informal labor). Each of these instances is characterized by producing something or someone as a "resource" to be harnessed in the production of a "good," be it family livelihood or intensive farming. In producing land, water, and people as a necessary "resource" of a particular kind, social agents authorize themselves to productively incorporate concrete biophysical entities into circuits of production. In the case of commercial farming, this is also a circuit of capital valorization.

In Huelva, the predatory meaning seems to crucially depend on the intervention of state institutions at various scales that legally designate what is predatory (i.e. illegal), that is, "productive *destruction*"—a violent incorporation of value—as distinct from what is regular "productive

[14] https://www.juntadeandalucia.es/sites/default/files/inline-files/2023/11/Encuesta%20de%20Precios%20de%20la%20Tierra%20en%20Andaluc%C3%ADa%202022_v1.pdf

consumption," capital on the move. Yet, the recurrent tolerance or regularization of the use of illegal resources, such as particular instances of groundwater extraction, land occupation, and forms of irregular labor force, points to the tenuous grounds on which the concept of "predation" as distinct from capitalist production more generally stands. Expelling resources from legal access immediately devalues them while it does not prevent their use. This appears as an internal process of devaluation akin to what Harvey (2006) defines as accumulation by dispossession. Through the active process of producing legal boundaries around resources such as water and labor, government administrations devalue those resources defined as "illegal" even as they tolerate their appropriation for capital accumulation. This legal infrastructure creates an internal commodity frontier that supports extractivism. As a result, productive destruction and its predatory aspect blurs into productive consumption, the ordinary process of transforming the value of a bundle of commodities into more valuable commodities. In Huelva, unlike in Latin America, while the state is a central player of the extractivist process in agriculture, its distributive intervention only helps transfer wealth to capital.

Indeed, the control and regulation of water—through material and legal infrastructures—is unequal and produces conflict at various scales, often around the valuation of water uses (Loftus, 2015; del Moral Ituarte et al., 2020; Walsh, 2022) and its quantitative (abundance, depletion) or qualitative (pollution, salinization) characteristics. Like labor, during production, water—and its value—is incorporated in commodities and accrues to capital. Tony Allan (1998) defines as "virtual water" the local water incorporated in fruits and vegetables for export that are consumed elsewhere. This hidden water extracted through export agriculture underscores unequal exchange between regions (Zlolniski, 2011; Hastrup, 2013; Velazquez, 2007). In the case of Huelva, value extraction incorporates virtual water which is depleted locally, hence a fundamental resource of intensive berry agriculture cannot be reproduced, a process of "production without reproduction" (Ye et al., 2019, p. 3).

Similarly, labor employed in food production has been displaced (e.g. through temporary migrant labor programs) and is disconnected from its site of reproduction (Burawoy, 1976; Prebisch & Binford, 2007). The widespread employment of migrant labor in food production all over Europe (Palumbo & Corrado, 2020; Pedreño Cánovas, 1999) foregrounds racialized, ethnicized, and gendered workers whose lives are systematically devalued in present-day advanced capitalist societies. In

Huelva, in addition to the separation of reproduction from production through migrant programs, maintenance of the labor force's immediate livelihood is constantly jeopardized and endangered. Reproduction is mostly unaccounted for, displaced to marginal spaces, and obscured.

In the case of intensive agriculture, the unaccounted use of water, solar radiation, and overexploited labor force become embodied in the produce exported and express an asymmetrical transfer of biophysical resources whose reproduction is not accounted for in market exchange, hence generating surplus value (Hornborg, 2019). It points, again, to the predatory aspect of this capitalist agriculture where productive destruction replaces productive consumption. This can be defined as extraction, as the use value of partially unpaid resources is accumulated in agricultural and other forms of capital. "Production without reproduction" comes to the forefront of what agricultural extractivism entails: (1) the active production of biophysical entities as "resources" that can be channeled into the circuits of capital valorization; (2) the constant play between legal/illegal/legitimate characterizations of relations between human and non-human resources and production (of commodities but also of a livelihood); and (3) discounting the processes of reproduction of some "resources" (water, labor), and materially and ideologically expelling them from production and valorization accounting.

Alf Hornborg (2019) builds on the exchange asymmetries between core and periphery and on the differentiation between biophysical resources and economic market valuation to propose a challenging theory of capital accumulation. His argument extends the central aspect of Marx's labor theory of value to other resources necessary to production such as land, energy, and raw materials. In the classic theory of value, surplus value arises from the discrepancy between an *economic asset*—labor power—which is exchanged at market value, and a *biophysical property*—labor—which is the capacity of human energy to contribute to the production of commodities with a value exceeding the socially necessary labor invested in the means of production. Hence the use value of labor power (i.e. a biophysical property) contributes more to the value of a commodity than its own exchange value (i.e. a market agreement).

But the point Hornborg wishes to make is that this is also the case for other resources, and that the asymmetrical transfer of resources' biophysical potential is concealed by the "fictive reciprocity of market prices" (Hornborg, 2019, p. 96). Hornborg bases his theory of value accumulation in an *internal* property of capitalism, namely the distinction "between the *biophysical aspects* of world trade (…) and the *economic categories* through which it is represented" (Hornborg, 2019, p. 154, our emphasis). Asymmetrical transfers of resources become central to the accumulation of value both through the extractive aspect of unequal exchange, as well as through the sociotechnical affordances that will enhance productivity and hence exploitation.

Should we understand this extractivism as predatory accumulation and as productive destruction? It could be argued that the incorporation of biophysical and sociocultural properties is the condition of capital's survival. Capital then could be understood as being simultaneously in a predatory and an exchange relational scheme with its human and non-human resources. Philippe Bourgois' rendering of predatory accumulation seems to closely relate to Harvey's idea of devaluing human and non-human nature in order to harness them productively to the circuits of capital valorization. In the case we have analyzed, the use of illegality—that Bourgois stresses—as an instrument in the process of accumulation underscores the crucial involvement of regulatory frameworks in the production of under- or de-valued resources. Illegality in the case of intensive strawberry farming in Huelva also becomes the argument that shifts land, water, or labor use from an exchange to a predatory regime.

However, as we have proposed in the previous section, Harvey's (2006) and Hornborg's (2019) frameworks seem to better explain the case. Both in Harvey's rendering of accumulation by dispossession and in Hornborg's idea of asymmetrical transfers of biophysical properties, a crucial aspect is the connection between production and reproduction, namely, the fact that reproduction of the material or energy used in production is not or only partially accounted for in exchange. We wish here to recall the twin concepts of "commodity fetishism" (Marx, 1982) and "fictitious commodities" (Polanyi, 1971), and point that in both cases things which appear to be identified by their position in market exchange (as commodities) are in reality dependent on the social relations that produce them at least partially outside market exchange. By looking at reproduction in capitalist production we find that what is extracted and accumulated, in one form or another, is always life.

Acknowledgments The ethnography in Andalusia benefited from the advice and collaboration of Alicia Reigada (Universidad de Sevilla) who has conducted fieldwork in that area starting in 2005, we wish to thank her for her invaluable help.

Competing Interests Natalia Buier's research project was supported by the European Union's Horizon 2020 research and innovation program under the Marie Skłodowska-Curie grant agreement No 890611 - ORIGINSOFSCARCITY - H2020-MSCA-IF-2019. Susana Narotzky's Research has been supported by grant PID2020-114317GB-I00 from the Spanish State Research Agency and the ICREA-Academia Fellowship Program, Generalitat de Catalunya.

References

Acosta, A. (2013). Extractivism and neo-extractivism: Two sides of the same curse. *Beyond Development: Alternative Visions from Latina America, 1*, 61–86.

Allan, J. A. (1998). 'Virtual water': An essential element in stabilizing the political economies of the Middle East. *Yale University Forestry & Environmental Studies Bulletin, 103*, 141–149.

Alonso-Fradejas, A. (2021). Leaving no one unscathed in sustainability transitions: The life purging agro-extractivism of corporate renewables. *Journal of Rural Studies, 81*, 127–138.

Alonso-Fradejas, A. (2024). The rise of authoritarian corporate populism. *Latin American Perspectives, 51*(6), 32–50.

Artiga-Purcell, J. A. (2022). Hydrosocial extractive territories: Gold, sugarcane and contested water politics in El Salvador. *Geoforum, 131*, 93–104.

Banerjee, S. B. (2008). Necrocapitalism. *Organization Studies, 29*(12), 1541–1563.

Barciela, C. (1996). La contrarreforma agraria y la política de colonización del primer franquismo, 1936–1959. In A. García Sanz & J. Sanz Fernández (Eds.), *Reformas y políticas agrarias en la historia de España* (pp. 351–398). Ministerio de Agricultura, Pesca y Alimentación.

Borras, Jr, S. M., Franco, J. C., Isakson, R., Levidow, L. & Vervest, P. (2014). Towards understanding the politics of flex crops and commodities: Implications for research and policy advocacy. *Think Piece Series on Flex Crops & Commodities, 1*, 1–15. Transnational Institute (TNI) Agrarian Justice Program.

Bourgois, P. (2018). Decolonising drug studies in an era of predatory accumulation. *Third World Quarterly, 39*(2), 385–398.

Burawoy, M. (1976). The functions and reproduction of migrant labor: Comparative material from Southern Africa and the United States. *American Journal of Sociology, 81*(5), 1050–1087.

del Moral Ituarte, L., Laconi, C., & Pedregal, B. (2020). Cartografiando el movimiento de justicia ambiental a escala regional: el mapa digital colaborativo de los conflictos del agua en Andalucía. *Boletín De La Asociación De Geógrafos Españoles, 85*, 1–35.

Filigrana, P. (2021). Del campo a los cuidados. El sindicalismo feminista y antirracista que viene. *Cuadernos #feminsimosindicalista3*, Laboratoria Espacios de Investigación Feminista, Madrid.
García Ruiz, T. (2018). Reflexiones sobre el estado actual de la técnica de construcción de pozos de captación de agua subterránea en España. Congreso Ibérico sobre Agua Subterránea, Medio Ambiente, Salud y Patrimonio. Salamanca, 12–15 November 2018, Universidad de Salamanca.
Gómez Benito, C. (1994). Joaquín Costa resituado: populismo, tradición campesina y materialismo hidráulico como definidores de su pensamiento social agrario. *Anales de la Fundación Joaquín Costa, 11*, 7–22.
Gualda Caballero, E., & Ruiz García, M. (2004). Migración Femenina de Europa Del Este y Mercado de Trabajo Agrícola En La Provincia de Huelva, España. *Migraciones Internacionales, 2*(4), 37–65.
Gualda, E. (2008). De Polacas a Senegalesas: Contratos En Origen, Sustitución Étnica y Segmentación Del Mercado Agrícola Onubense. *Panorámica Social, 8*, 71–84.
Gualda, E. (2012). Migración circular en tiempos de crisis. Mujeres de Europa del Este y africanas en la agricultura de Huelva. *Papers: revista de sociología, 97*(3), 613–640.
Gudynas, E. (2010). La ecología política de la crisis global y los límites del capitalismo benévolo. *Íconos-Revista de Ciencias Sociales, 36*, 53–67.
Harvey, D. (1999 [1982]). *The limits to capital.* Verso.
Harvey, D. (2003). *The new imperialism.* Oxford University Press.
Harvey, D. (2006). Neo-liberalism as creative destruction. *Geografiska Annaler: Series B, Human Geography, 88*(2), 145–158.
Hastrup, K. (2013). Water and the configuration of social worlds: An anthropological perspective. *Journal of Water Resource and Protection, 5*(04), 59–66.
Hellio, E., & Moreno Nieto, J. (2021). La Ecología-Mundo Bajo Plástico: Un Análisis de La Articulación Entre La Explotación de La Naturaleza, El Racismo y El Sexismo En La Producción de Frutos Rojos de Huelva. *Relaciones Internacionales, 47*, 125–142.
Hornborg, A. (2019). *Nature, society, and justice in the anthropocene. Unraveling the money-energy-technology complex.* Cambridge University Press.
Jiménez Díaz, M. (1992). La colonización agraria en la provincia de Huelva: los procesos de transformación del Avitorejo. *Huelva en su Historia, 4*, 379–390.
Loftus, A. (2015). Water (in)security: Securing the right to water. *The Geographical Journal, 181*(4), 350–356.
Márquez Domínguez, J. A. (1986). *La Nueva Agricultura Onubense: Protagonismo Agrario Del Municipio de Moguer.* Instituto de Desarrollo Regional, Universidad de Sevilla.
Márquez Domínguez, J. A., Gordo Márquez, M., Felicidades García, J., & Díaz Diego, J. (2013). Evolución de los contingentes de trabajadores extranjeros en

España (1993–2011): las actividades agrícolas. In J. A. Camacho Ballesta & Y. Jiménez Olivencia (Eds.), *Desarrollo Regional Sostenible en tiempos de crisis* (Vol. 2, pp. 757–780). Universidad de Granada.

Marx, K. (1982). *Capital. A critique of political economy* (Vol. 1). Penguin Books / New Left Review.

Marx, K. (1993 [written 1857–8, first published 1939]). *Grundrisse. Foundations of the critique of political economy (rough draft)* (M. Nicolaus, Trans.). Penguin Books / New Left Review

Mbembe, A. (2003). Necropolitics. Translated by Libby Meintjes. *Public Culture, 15*(1), 11–40.

McKay, B. M., Alonso-Fradejas, A., & Ezquerro-Cañete, A. (2021). Introduction. In B. M. McKay, A. Alonso-Fradejas, & A. Ezquerro-Cañete (Eds.), *Agrarian extractivism in Latin America*. Routledge.

Molinero Gerbeau, Y. (2020). Dos décadas desplazando trabajadores extranjeros al campo español: una revisión del mecanismo de contratación en origen. *Panorama Social, 31*, 141–153.

Montag, W. (2005). Necro-economics. Adam Smith and death in the life of the universal. *Radical Philosophy, 134*, 7–17.

Murphy, M. (2017). *The economization of life*. Duke University Press.

Narotzky, S. (2021). Agricultural day labor in Spain: The logics of (pandemic) capitalism. In D. Fassin & M. Fourcade (Eds.), *Pandemic exposures: Economy and society in the time of Coronavirus* (pp. 247–273). HAU / The University of Chicago Press.

Neocleous, M. (2003). The political economy of the dead: Marx's vampires. *History of Political Thought, 24*(4), 668–684.

Ojeda Rivera, J. F. (2022 [1987]). *Organización del territorio en Doñana y su entorno próximo (Almonte)*. Editorial Niebla.

Ortega Cantero, N. (1984). Las propuestas hidráulicas del reformismo republicano: del fomento del regadío a la articulación del Plan Nacional de Obras Hidráulicas. *Agricultura y sociedad, 32*, 109–152.

Ortí, A. (1984). Política hidráulica y cuestión social: orígenes, etapas y significados del Regeneracionismo Hidráulico de Joaquín Costa. *Agricultura y sociedad, 32*, 11–107.

Palumbo, L., & Corrado, A. (Eds.). (2020). *Are Agri-food workers only exploited in Southern Europe? Case Studies on Migrant Labour in Germany, the Netherlands, and Sweden*. Open Society Foundations.

Pedreño Cánovas, A. (1999). Construyendo la 'huerta de Europa'. Trabajadores sin ciudadanía y nómadas permanentes en la huerta murciana. *Migraciones, 5*, 87–120.

Petras, J., & Veltmeyer, H. (2014). *Extractive imperialism in the Americas. Capitalism's New Frontier*. Brill.

Polanyi, K. (1971 [1944)]). *The great transformation*. Beacon Press.

Prebisch, K., & Binford, L. (2007). Interrogating Racialized Labour Supply: An Exploration of the Racial/ National Replacement of Foreign Agricultural Workers in Canada. *Canadian Review of Sociology, 44*(1), 5–36

Reigada, A. (2022a). *Historia, trabajo y territorio. El conflicto capital-vida en los campos de fresas de Huelva*. Universitat de Barcelona Edicions.

Reigada, A. (2022b). A link in global agrifood chains: Recruitment policies, work, and sexuality in the strawberry fields of Andalusia (Spain). *Current Anthropology, 63*(5), 519–540.

Schumpeter, J. A. (1994 [1942]). *Capitalism, socialism and democracy*. Routledge.

Svampa, M. (2019). *Neo-extractivism in Latin America: Socio-environmental conflicts, the territorial turn, and new political narratives*. Cambridge University Press.

Swyngedouw, E. (2015). *Liquid power: Contested hydro-modernities in twentieth-century Spain*. MIT Press.

Velázquez, E. (2007). Water trade in Andalusia. Virtual water: An alternative way to manage water use. *Ecological Economics, 63*(1), 201–208.

Walsh, C. (2022). Beyond rules and norms: Heterogeneity, ubiquity, and visibility of groundwaters. *Wiley Interdisciplinary Reviews: Water, 9*(4), e1597.

Ye, J., Van der Ploeg, J. D., Schneider, S., & Shanin, T. (2019). The incursions of extractivism: Moving from dispersed places to global capitalism. *The Journal of Peasant Studies, 47*(1), 155–183.

Zlolniski, C. (2011). Water flowing north of the border: Export agriculture and water politics in a rural community in Baja California. *Cultural Anthropology, 26*(4), 565–588.

"Non-profit" Intermediation? Housing Access for Migrant Farmworkers in Italy Amidst the Criminalisation of *Caporalato*

Cecilia Vergnano

Introduction

Media representations, and some scholarly ones, of living and working conditions of migrant farmworkers employed in Italian industrial agriculture tend to focus on two elements, presenting them as characteristics of the Italian model of labour organization in the agro-industrial sector: (1) farmworkers' appalling housing conditions in makeshift slums (labelled as "ghettos" by dwellers and outsiders alike) and (2) labour intermediation through *caporalato* (gangmastering—further discussed below). Importantly, c*aporali* (gangmasters) usually provide a range of services to farmworkers which go beyond mere labour intermediation and include (substandard) housing and transportation to the workplace—according to a model of labour organisation where productive and reproductive spheres

C. Vergnano (✉)
Université Libre de Bruxelles, Group of Research on Ethnic Relations and Equality (GERME), Bruxelles, Belgium
e-mail: cecilia.vergnano@ulb.be

S. Narotzky et al. (eds.), *Agricultural Extractivism in the Mediterranean Region*,
https://doi.org/10.1007/978-3-032-05599-6_6

are strictly entangled. It is important to notice that these two elements (ghettos and *caporali*) tend to be represented, at least in some narratives, as interrelated: farmworkers' makeshift slums are often described as the condition of possibility for *caporali* to exert their power (as "logistical bases used to subjugate the immigrants", as expressed by Fanizza & Omizzolo, 2018, p. 3).

While critical scholars have questioned the centrality of the role of *caporali* in the exploitation of farmworkers, the representation of *caporalato* as the *cause* of exploitation (rather than its consequence) has ultimately shaped policy responses in recent years. By diverting attention from the structural character of labour exploitation in the neoliberal corporate food regime (McMichael, 2009), Italian state responses to eradicate labour exploitation in the agro-industrial sectors have indeed mostly focused on the criminalisation of *caporalato*. Such a focus is merely based on a criminal justice approach, which leaves the structural conditions of labour exploitation (such as the concentration of retail power and the increased competition resulting from trade liberalisation in food products at the global level) untouched (Caruso & Macciani, 2023; Dines, 2023). Furthermore, as farmworkers' informal settlements are identified as the condition of possibility for gangmasters' abuses, state efforts to eradicate *caporalato* often go hand in hand with the eradication of informal settlements—evictions of farmworkers' ghettos are indeed justified precisely as a means to prevent illegalities.

Historically rooted in the Italian southern countryside, *caporalato* is a model of labour intermediation in which some brokers or intermediaries, called *caporali*—not necessarily linked to organised crime—act as a link between labour demand and labour supply, that is to say, between farmers and labourers, in return for a fee from both parties—as the (informal) equivalent of a contemporary job agency. Importantly, in contrast with the inefficiency of public employment offices and the relatively higher costs of private job agencies, in many Italian agricultural districts, *caporalato* might be "*de facto* the only system of work mediation to efficiently guarantee the quota of just-in-time availability of unskilled labour and ensure lower labour costs" (Corrado et al., 2018, p. 3). As a result of a series of processes affecting the whole agro-industrial sector, such as globalisation and ethnic segmentation of the labour markets, *caporali* tend to be of the same nationality or ethnicity as the workers they manage: just as local labourers have been replaced by migrant workers, Italian *caporali* have been replaced by migrant *caporali*. Typically, a *caporale* is a former farmworker; in many instances, he lives in the same settlement with the workers, which allows him to continue to manage his workforce also during non-working hours (thus contributing to farmworkers' seclusion as the *caporale* might prevent or control their interactions with outsiders: journalists, researchers, other *caporali* offering better working conditions, etc.).

Literature suggests that *caporali*'s role is de facto ambivalent, oscillating between exploiter and workers' advocate: for instance, a *caporale* can lower the price of piecework to offer more attractive conditions to farmers (a strategy usually adopted to compete with other *caporali* and take over their markets) or, to the contrary, he can support workers acting as their spokesperson when working conditions become too exploitative (Caruso & Macciani, 2023; Perrotta, 2014). Furthermore, as one of the central pillars of critical criminology maintains, informal economy does not necessarily mean *criminal* economy. However, in recent decades, similarly to the figure of the migrant smuggler, the figure of the *caporale* has been subjected to a process of criminalisation. This reflects the so-called "modern slavery paradigm" often underlying policy and media representations, which tend to portray the *caporale* as the villain, migrant farmworkers as victims without agency, and farmworkers' informal settlements as hubs of illegality (Caruso & Macciani, 2023). On the ground of such an oversimplified representation, two laws (in 2011 and 2016) were issued by the Italian government to eradicate *caporalato*.[1]

How do migrant farmworkers access housing in a context where labour intermediation—which, as mentioned, often goes hand in hand with *housing* intermediation—is criminalised, and the emergence of new informal settlements is precluded? By focusing on the case study of an important agricultural district in southern Italy, this chapter explores continuities and transformations in patterns of housing access for racially subaltern farmworkers following the criminalisation of labour intermediation in Italy. Empirical data collected in the field reveal the formation of specific "arrival infrastructures" (Meeus et al., 2019) and the emergence of new intermediaries, including actors from the humanitarian sector, the third sector and local residents, who have adopted a utilitarian logic in their relationship

[1] In 2011, a new type of offence was introduced in the Italian Criminal Code, under the name of "unlawful intermediation and labour exploitation" (also called "article 630-bis"), which exclusively punishes the intermediaries, thus leaving the employers untouched. Following that, in 2016, the Italian parliament approved the more comprehensive law 199/2016, under which *both* labour intermediaries and employers engaging with them became prosecutable.

with migrant workers—being such logic part and parcel of hegemonic political rationalities in migration management (both progressive and reactionary ones).

The chapter is structured as follows. I will first present a conceptual framework based on two main concepts: (1) arrival infrastructures, and (2) refugeeisation of agricultural workforce in the Italian context. I will then analyse the discursive context in which research is conducted, characterised by two main competing (but complementary) political rationalities in contemporary migration politics, which can be schematically labelled as (neo)liberal migration management, on the one side, and securitisation/criminalisation of migration, on the other. By focusing on a specific agricultural district in southern Italy, whose name has been pseudonymised for ethical purposes (as I will detail below), the chapter then analyses continuities and transformations in housing access for racially subaltern farmworkers in the aftermath of a paradigmatic event, that is, the eviction of a farmworkers' ghetto, an informal settlement in an abandoned industrial area, which was home in recent years to several hundreds of migrant farmworkers.

Importantly, because the ethnographic material contained in this chapter might result in legal liability for some research participants, I adopted fictional narrative techniques. The name of the agricultural district, Grande Piana, is fictional, as are the names of the two localities (fictionally called San Lazzaro and San Giulio), while the names of research participants are pseudonymised as usual in anthropological research. Chronological references are also purposely vague to avoid the traceability of specific events. The only names maintained are those of two important organisations committed to peasant agriculture, food sovereignty and fair trade: the peasant union Altragricoltura and the association No CAP. As the two organisations are active at the national level, their identification does not make research participants in this particular field recognisable.

Research findings are based on a four-month fieldwork between 2023 and 2024, complemented by in-person interviews, digital ethnography, and analysis of digital content to retrieve data on those events preceding the fieldwork (such as the eviction of the black farmworkers' settlement analysed below). Interviewees include migrant farmworkers and other residents, including actors taking part, at the local level, in the No CAP project (explained below).

Arrival Infrastructures in Rural Areas and the *Refugeeisation* of the Agricultural Workforce

The concept of migration infrastructure was first proposed by migration scholars Xiang and Lindquist in their influential work (2014) to indicate the intertwined set of technologies, institutions, and actors that facilitate and condition human mobility. Migration infrastructure is mainly applied to labour migration and encompasses five dimensions: (a) commercial (e.g. labour intermediaries), (b) normative (governmental mechanisms and procedures, etc.), (c) technological (communication and transport), (d) humanitarian (NGOs and international organisations), and (e) social (e.g. migrant networks). While these five aspects might conflict and contradict each other, their interconnection is key to understanding migration infrastructure.

Building on the aforementioned paradigm of migration as a social process that takes place through a specific infrastructure, another group of scholars (Meeus et al., 2019) propose the concept of *arrival* infrastructure. By arrival infrastructure, the three authors propose a notion of "arrival" understood not as the final goal of the migratory career from a spatial, temporal, and, therefore, teleological perspective, but rather as a kind of "landing platform" that provides migrants with some kind of stability to negotiate and build their social mobility (either on the place or through further translocal mobility).

While the concept of arrival infrastructure has been mostly developed through research in urban contexts, a focus on arrival infrastructures in rural contexts presents several potentialities to broaden and enrich the debate. Indeed, many rural areas in Europe have become superdiverse while at the same time lacking infrastructure (transport, communication, schools, hospitals) in a way that affects all residents, autochthonous and foreign migrants alike (Papadopoulos & Fratsea, 2017; Schneider, 2022; Wessendorf & Phillimore, 2019). This creates a specific framework of opportunities and limitations for migrants arriving in rural areas, while their very arrival, in turn, creates opportunities for locals but also reactions of moral panic, in a context characterised by structural racism and decades-long processes of depopulation, as I will detail below.

While Italy shifted from being a primarily emigration country to a net immigration country in the 1980s, the net difference between emigration and immigration rates has declined in recent years, with fewer foreign citizens entering and more Italians leaving than before. This phenomenon is particularly visible in rural areas of the south, where the local population is dwindling and ageing, and newcomers are mostly foreign immigrants. In a context of neoliberal globalisation, where the agricultural sector has progressively adopted the industrial Californian mode, the availability of migrant workforce (relatively more docile and controllable than national one, due to migrants' subaltern legal status) in agricultural districts of the Italian south has allowed some small and medium farms to remain relatively competitive in the global market even in the absence of technological investments (which they could not have afforded). In other words, while in agricultural districts in the rest of Europe (as well as northern Italy) the trend in mechanisation and corporate concentration is more marked, in southern Italy, the availability of cheap migrant labour has provided a viable alternative to (unaffordable) mechanisation for small and medium farms, thus contributing to delaying their disappearance (Perrotta, 2016).

In Italy, furthermore, the management of farmworkers has been influenced in the last decades by specific policies and interventions, characterised by an underlying logic of *refugeeisation* of the agricultural workforce. Originally proposed by Nick Dines and Enrica Rigo (2015), the concept of *refugeeisation* accounts for the penetration of the "humanitarian reason" (Fassin, 2012) into the management of this specific sector of the workforce. Importantly, the concept of *refugeeisation* does not merely refer to the fact that migrant farmworkers' legal status is increasingly granted based on humanitarian protection rather than for work reasons (although this is certainly true, especially for a key component of agricultural labour, that is, Western African workers who arrived to Italy as asylum seekers), but rather to the tendency, by state and civil society actors, to *represent* and *manage* racially subaltern migrant workers as (potential) *victims* of abuse in need of protection and surveillance (Dines, 2018), irrespective of whether they are actual refugees or not—and regardless of their actual legal status. Be they asylum seekers, holders of humanitarian protection, or granted a residence permit on other legal basis, racially subaltern migrant farmworkers tend to be represented, in mainstream media and political discourses, as *victims* (of the *caporali*, of human traffickers, etc.) and therefore subject to humanitarian interventions rather than actual labour policies. Their identity as workers is no longer central, thus allowing a discursive shift towards the language of deservingness and emergency: they deserve reception rather than housing; camps rather than proper accommodation; caritative assistance rather than social and political rights in all respects. This includes, among other things, their placement in container camps run by civil protection forces (or humanitarian or religious organisations) (Rigo & Dines, 2016). Container camps are presented, in turn, as interventions to combat the illegal phenomenon of *caporalato* (Caruso & Corrado, 2021).

Political Rationalities in Migration Management: Migrants as a Threat "vs." Migrants as a Resource

Migration governance is currently characterised by two competing but complementary (rather than mutually exclusive) discursive fields where migration is presented as either a threat or a resource; a security problem or an economic opportunity, a potentially threatening phenomenon which requires restrictions, or a potentially positive phenomenon that has to be correctly managed to maximise its positive outcomes. Following a Foucauldian analytics of government (which is concerned with the conditions under which certain governing practices emerge and appear rational), "political rationality" can be defined as a particular although not necessarily homogeneous discursive field within which the exercise of power is conceptualised, and where forms, objects, and limits of politics tend to emerge as self-evident (Rose & Miller, 2010). As it concerns the discursive field of migration, different political rationalities can be schematised as the restrictive or securitarian approach, on the one side, and the migration-management approach, on the other. Importantly, "political rationalities are heterogeneous, overlapping and often transformed as a result of the shifting constellations of related practices and mentalities of governing existing in the surrounding policy landscape" (Kalm, 2010, p. 23). Among the "shifting constellations" surrounding migration discourses, it is important to mention neoliberal governmentality and humanitarian reason. The latter is, as already mentioned, pivotal in shaping mainstream representations of migrant farmworkers—as well as their management—as a key feature of the already-mentioned process of *refugeeisation* of the agricultural workforce.

While states are still the paramount actors in governing migration, a range of international organisations and forums now exert influence in this area, such as the International Organisation for Migration (IOM), the Global Commission on International Migration (GCIM), and the Global

Forum for Migration and Development (GFMD). Historically, the (neo) liberal migration management approach has emerged as a discourse promoted precisely by these global organisations and forums, in a historical conjuncture (at the end of the 1990s) when the imperatives of neoliberal globalisation stood out prominently.

The basic pillars of the migration-management approach, as proposed by these global organisations and forums, are that: (1) international migration is a "natural" phenomenon (rather than being driven by periphery/core inequalities within the world-system: see Balibar & Wallerstein, 1988), and (2) migrants contribute positively to the economies of receiving societies. Currently, such a political rationality is often implicit in the discourses of (some) progressive Western political leaders, humanitarian professionals, members of radical social movements committed to social justice issues, and representatives of civil-society organisations within the galaxy of so-called "refugee welcome" movements.[2]

At first glance, the emergence of the migration-management discourse, which stresses the positive, win-win impacts of migration (if well managed) on both sending and receiving societies, might appear as "a breath of fresh air" (Geiger & Pécoud, 2010) compared with the securitarian approach, which promotes a "draconian" governance of migration (De Genova, 2013; Kalir, 2019). However, as well illustrated by a number of scholars (De Genova, 2013; Grappi, 2020; Kalm, 2010), the migration-management discourse underlies an instrumental, technocratic, and depoliticising approach towards migration. Indeed, even when migration is not conceived as a threat to the welfare and security of receiving states, it implies that migration still needs to be guided in a top-down manner and according to the demands of receiving states—to be more precise, according to *employers'* demand, thus producing "a flexible labour force [...] prepared to work for low wages and under difficult conditions" (GCIM, 2005, p. 6).

Importantly, the two political rationalities should not be seen as incompatible with each other but rather as the two sides of the same coin: the securitarian or restrictive approach is functional to labour subordination in

[2] Examples of it are the discourse of Spanish liberal left party (PSOE) (https://tinyurl.com/ktwkx6rt), or the commitment of the German "refugee welcome" movement to facilitating the match between German employers' demand and the offer of "unexpected labour power" provided by asylum seekers in the aftermath of the so-called EU "refugee crisis", while German employers' representatives were literally quoting No Borders' claims for granting asylum seekers access to the labour market (Altenried et al., 2018; see also Maaroufi, 2017).

labour relations (De Genova, 2013) which, in turn, is functional to maximise the "positive" outcomes of migration for receiving societies (read capital accumulation through migrant labour exploitation), as envisaged by the migration-management approach.

The Agricultural Area of *Grande Piana*

The Eviction of La Nuova Italia Industriale (2019)

While it is always difficult to establish when a social or historical process starts, as the very decision to "narrate a story" from a particular moment or event—rather than another—influences the narrative itself (Mwambari, 2023), the "story" I recount in this chapter starts a few years ago with the eviction of a settlement of black farmworkers in an abandoned industrial building, which I will fictionally call La Nuova Italia Industriale, on the outskirts of the village of San Giulio in the Grande Piana. In this area, in recent decades—just as in many other Italian agricultural districts—agricultural production went through a deseasonalisation process to increase productivity. Consequently, the demand for workforce (although still fluctuating throughout the year) is relatively stable. In this context, the settlement of La Nuova Italia Industriale played an important role as an arrival infrastructure, providing housing to hundreds of workers, mostly from Western Africa and mostly working on local farms—including three *caporali* (from Libya, Burkina Faso, and Gambia) who lived in the settlement. Importantly, access to La Nuova Italia was regulated by the *caporali* themselves. As usual in this model of labour intermediation, they provided dwellers with accommodation in shacks built within the industrial area, as well as transport to their workplaces, in return for fees (besides providing work).

The settlement of La Nuova Italia Industriale was neither the first nor the last in the area, as I will detail below. Nevertheless, its eviction can be seen as a turning point in local governance, in that the eviction (unlike previous evictions of other *grand ghettos* in the area) was *not* followed by the emergence of a new informal ghetto of the same dimensions—hence my decision to set the beginning of the narrative here. The demand for cheap labour in the local agro-industrial sector, although subject to fluctuations, did not substantially decrease after the eviction of La Nuova Italia Industriale. Therefore, the disappearance of *grand ghettos* in the area cannot be attributed exclusively to black workers' displacements to other agricultural districts. Rather, the dwellers' spatial dispersal was followed by the emergence of new arrival infrastructures, as I will detail below.

For the sake of accuracy, it is important to remark that occupations of abandoned buildings in the area still take place (and are the object of systematic evictions), and shacks in abandoned plots of land continue to be constructed (and systematically demolished by public order authorities). However, the eviction of La Nuova Italia put an end to the age of *large ghettos* providing housing to hundreds of dwellers, and inaugurated a new phase in which, at most, only small, invisible settlements of a few individuals are tolerated, and new arrival infrastructures have emerged. To understand how this occurred, it is necessary to take a step back to the anti-eviction mobilisations, preceding and following the eviction itself.

In such mobilisations, a prominent role was played by a constellation of organisations (mostly national rather than locally based) promoting different models of agricultural production through fair trade, peasant agriculture, and food sovereignty, gathered under the umbrella of Forum di Dignità (Dignity Forum). The Forum included a farmworkers' committee composed of the most politicised ghetto dwellers, the association No CAP (a name which indicates the main goal of the association: the eradication of *caporalato*), and Altragricoltura (a peasant union struggling for food sovereignty against globalised agriculture), among other actors.

Among the actions organised by the Forum, a demonstration of farmworkers took place after public authorities' announcement of the imminent eviction of La Nuova Italia. This was followed by a public assembly. During the demonstration, a group of farmworkers spontaneously started a roadblock. However, according to the testimonies I collected, leaders of Altragricoltura and No CAP put pressure on them to desist, carrying out intermediation with authorities and adopting an attitude that some of the participants defined as paternalistic.

The protest did not succeed in stopping the eviction or forcing local authorities to provide viable housing alternatives to the dwellers. Yet, it is remarkable that, for the first time, an eviction of a ghetto of black farmworkers achieved such a level of politicisation at the local level (even though, as mentioned, the Forum was mostly composed of non-local organisations). At the same time, such a politicisation was based on a specific representation of migrant farmworkers' exploitation: one in which *caporalato* was conceived as the primary cause of exploitation (rather than one of its effects), and legality was framed as the main solution to it - thus downplaying the structural causes of exploitation.

In the aftermath of the eviction, new arrival infrastructures emerged in the area. Some arose on the initiative of the local Catholic Church, in connection with No CAP and Altragricultura and their networks at the local level; others, by contrast, arose on the initiative of local landlords and local holders of rental contracts (both Italian and foreigners, the latter including the same *caporali* previously operating at La Nuova Italia Industriale).

Non-profit Intermediation? Casa della Dignità and the No CAP Project

Before and after the eviction of La Nuova Italia Industriale, the Forum di Dignità repeatedly demanded alternative plans for the reception of migrant farmworkers, underlining that it was local authorities' responsibility to provide reception. I use the word "reception" (*accoglienza*, in Italian) rather than "housing" not by mistake, but because it was the word used by the most prominent spokespersons of the Forum. This choice once again reveals the paternalistic attitude towards migrants within a part of the Forum itself—the part with more voice in the media—in a context of *refugeeisation* of agricultural workforce, as mentioned above. While it is true that farmworkers' housing needs might fluctuate due to seasonal variation in agricultural labour (which can partly explain why the leaders of the Forum di Dignità were demanding reception rather than housing), one could question whether "reception" represents a form of "dignity" for subjects who are effectively workers. One could question, as well, why public authorities (and not, for instance, private farmers who extract value from farmworkers' labour) are responsible for providing housing/reception. In fact, the very presence of Altragricultura (which is, it is worth reminding, a union of small agricultural entrepreneurs) within the Forum explains the erasure of a dimension of class in the conflicts following the eviction of La Nuova Italia; while a committee of farmworkers of La Nuova Italia was part of the Forum, their voice was significantly underrepresented in the media and tactical decisions were mostly taken by the other components of the Forum (Altragricoltura and No CAP).

As local authorities did not provide any answer to farmworkers' housing needs following the eviction of La Nuova Italia, other initiatives arose from private actors of the non-profit sector. One of these initiatives was Casa della Dignità ("House of Dignity"), a reception facility for migrant farmworkers funded by the Church, located in an isolated area a few kilometres from La Nuova Italia and the village of San Giulio. Casa della

Dignità could accommodate a small number of farmworkers (approximately twenty) who had at their disposal a common kitchen, a prayer room, and two rooms for job training purposes (Italian language, IT). One of the farmworkers, together with the diocesan director, was responsible for the management of the reception. As stated by the diocesan director, the place represented "an opportunity to leave behind inhuman and shameful conditions. To leave behind the risk of ending up in crime networks and *caporalato*", thus confirming the representation of *caporalato* as the cause of farmworkers' exploitation—rather than one of its effects.

On the day of the inauguration of Casa della Dignità, representatives of No CAP and Altragricoltura intervened with public speeches, together with local authorities. No CAP and Altragricoltura leaders were presenting their brand of ethical products distributed by an important retail chain and certified with the No CAP label, guaranteeing that no worker has been exploited by *caporali* during the harvest. Most of the farmworkers living in Casa della Dignità were indeed employed by a local farm in the framework of the No CAP project.

At this point, I should describe the No Cap project and the way it was implemented. The central idea of the No Cap project is to create a special label certifying that a particular agricultural product (fresh or transformed) has been produced without labour exploitation. In practice, the first step for the No CAP association is to propose to large retail chains their "No CAP" label. Once retailers have shown an interest, No CAP turns to farmers. It is the farmers who establish the fair price for specific products. Importantly, the fair price covers labour costs calculated according to legal provisions, in order to comply with existing labour regulations. In other words, it is guaranteed that farmworkers' salaries and working conditions comply with the law, nothing less and nothing more. Once large retailers confirm their orders, No CAP pass it to farmers. No CAP can also provide workers to farmers, as well as a service to transport them to the workplace. Furthermore, it performs controls throughout the whole production process. While No CAP does not directly provide accommodation to workers, it collaborates at the national level with existing projects provided by the third sector or the Catholic Church, such as Casa della Dignità.

As the reader can easily understand, the central hypothesis at the basis of the whole project is that if there is no *caporalato*, and legal conditions are met, there is no exploitation. Since its foundation in 2018, No CAP has received media coverage, being generally praised in media representations as a "good model" to eradicate labour exploitation in the agro-industrial sector. From a critical perspective, however, two elements of the project are particularly striking. The first one is the controversial character of a label, which ultimately certifies that the product complies with existing law. On the one side, it implicitly assumes that law fulfils its role in protecting workers (which is a problematic assumption); on the other, it insinuates that equivalent products *without* the No CAP label, often just next to No CAP products on supermarket shelves, have been produced in an illegal way.[3]

The second problematic element in the model proposed by No CAP is that, ultimately, No CAP itself becomes the intermediary or, in other words, No CAP takes on the central role (both in production and reproduction processes) previously assumed by *caporali*. As explained by the president of No CAP, Yvan Sagnet, a Cameroonian man who has been a victim himself of severe labour exploitation while working in the tomato harvest (and became in 2011 the leader of the largest strike of migrant farmworkers so far in Italy; he then made a career as a trade unionist until founding No CAP):

> Our goal as No-Cap is to *compete with* caporali *and crime*, taking the men and women they exploit, taking them out of their hands and putting them into the legal labour circuit we are building. *We find legal jobs for the workers, we provide accommodation, and we provide transportation.* (emphases added)

Based on his experience of living in Casa della Dignità and working in the framework of the No CAP project, a Western African farmworker, Amadou, draws the logical conclusion of the discourse of the No CAP president:

> No CAP is the new *caporale*. It is obvious: without *caporale* there's no job. This is a new form of colonialism. Who is paying the electricity bill to No CAP? It's me, with my job. […] I do not like to work legally. The salary [within the framework of the No-CAP project, that is to say, in full

[3] The communication officer of a large retailer which commercialises No CAP products acknowledges this problematic aspect when she says in an interview that "We are not allowed to say that the product with the No CAP label is a 100% ethical product and instead the neighboring product on the same shelf is a product of *caporalato*, because obviously we have no evidence or counter-evidence", just to somehow contradict herself immediately after by adding that "We are not in the position to change the rules of the game, but we can do something by creating our own line of certified products in which we put our face, and No CAP acts as a guarantor".

> compliance with the law] was only 42 euros per day, plus I had to pay 4 euros for transport. [...] I have just worked under a real *caporale* in the collection of watermelons: I got 1000 euros in two weeks. With No CAP I got 766 euros in 18 days, minus the transport. Put yourself in my shoes, make your math and tell me which one you would choose.

In light of Amadou's words, it becomes visible how the intermediation by No CAP ensures compliance with the law, but the law also allows exploitative conditions. Concerning the pretension of No CAP's president to replace *caporali*, or compete with them, some problematic aspects are worth mentioning. Indeed, while confirming the structural need for intermediation, No CAP is exposed to the risk of recreating the same conditions of workers' dependency characterising the *caporalato* model: in other words, No CAP risks reproducing a model where workers' both productive and reproductive needs are satisfied by one single broker and, therefore, power relations between the broker and the worker are, again, at play. Power unbalance is worsened by the fact that the whole No CAP project still relies on a certain degree of informality (read: arbitrariness) in its networks at the local level, as I will show below.[4]

The case of the transportation fee imposed on No CAP workers by Alessandro can serve as an example of such arbitrariness. Alessandro, an Italian man living in the village of San Lazzaro (not far from La Nuova Italia Industriale), is a local No Cap collaborator. In particular, he was responsible for driving the van to bring workers from Casa della Dignità to the farm where they were employed—a farm collaborating with the No CAP project. As he explained when I interviewed him,

> I believe migrants can contribute to our economy and mitigate the depopulation of our villages. I mean, we need workers and our villages are getting empty ... What more can we ask for than the arrival of migrants? Furthermore, migrants are much better than many Italians. I always say that they represent the best part of our society: they are law-abiding, honest, respectful, hard workers. They make our villages better places.

[4] I have personally collected denunciations against one of the collaborators of the project at the local level: the migrant farmworker who was in charge of the reception management of Casa della Dignità—accused by other residents of charging them with abusive fees in return for simple services such as going to the nearest village to buy cigarettes (it is worthy to remind that Casa della Dignità is spatially isolated, being located on a highway a few kms away from San Giulio), or helping them with applications for residence permits. Ultimately, No CAP has removed him from office.

While highlighting the win-win outcome of migrants' arrival in depopulated villages, an argument aligned with the migration-management discourse outlined above, Alessandro erases migrants' subjectivities, taking for granted, for instance, that life in rural areas in southern Italy is the ultimate aspiration for a migrant, rather than a constraint due to lack of better opportunities. He then goes on to explain how he positioned himself in the space between labour demand and labour supply at the local level:

> I created an association here in the village, precisely because I believe in another way of doing agriculture, and I believe that migrants can contribute so much to it. As an association, we collaborate with No-Cap: we bought a van and used it to bring the farmworkers to the farm involved in the No-Cap project. I was the van driver. I used to ask them [he refers to farmworkers] 4 euros for the round trip to cover the expense for the van.

Alessandro cannot be reductively labelled as an exploitative profiteer unscrupulously attempting to maximise his profit. Although not a farmer himself (he is simply embedded in a network of local farmers), he is an enthusiastic supporter of more ethical models of agricultural production and, more generally, is committed to the promotion and development of his village, against demographic ageing, depopulation, and electoral success of conservative, anti-immigrant parties in local administrations. Nevertheless, the request of a fee to workers (rather than the farmer) in return for transporting them to the workplace does not differ from a *caporale*'s request for the same service. Moreover, the justification offered -covering the expenses for the van- is questionable, since the van remains the property of the association, and not of the workers who contributed to its costs.

Since the inauguration of Casa della Dignità, Alessandro committed to participating in initiatives taking place there: "I used to go there in the evening bringing fresh products from local farms [for free], with the idea to organise collective dinners". Soon, however, he realised some limitations in the functioning of Casa della Dignità.

> The first problem is that forced cohabitation does not necessarily mean pleasant cohabitation. Most of the people in Casa della Dignità did not even show an interest in having dinner together. But the bigger problem was the lack of access to legal residence. People cannot register in the Municipality

with the address of Casa della Dignità, and this is a big problem to renew documents, etc.[5]

Moving from this acknowledgement, Alessandro started playing an important role in the emergence of new arrival infrastructures at the local level. As he was in touch with the landlords of San Lazzaro who owned uninhabited real estate, he started acting as an intermediary to facilitate rental contracts to farmworkers—most of them Western Africans and, as such, traditionally excluded from the local rental market due to racial prejudices. When I asked Alessandro whether it is thanks to him that people of his village started to rent their properties to *black* farmworkers for the first time, his answer was "Definitely yes". Importantly, the price for the rental is affordable for farmworkers, as these properties have virtually no value in a context of depopulation, where real estate assets in the village tend to remain empty.

Alessandro did not request any direct fee in return for his housing intermediation, as confirmed by the farmworkers I met in the fieldwork. For Alessandro, the farmworkers he has been able to accommodate in San Lazzaro represent a resource (even if he did not directly extract a profit from them) as they form a "pool of reliable labour", in the sense that he knows them personally and can guarantee they are trustworthy to those farmers in the area (especially those belonging to the farmer union Altragricoltura, in which Alessandro is involved as a militant) who are looking for seasonal labourers. In return for his intermediation, Alessandro asks farmers for their surplus of fresh produce,[6] which he then distributes throughout a local group (also founded by him) for ethical purchases.

The new presence of black residents in the village is not experienced without friction and tensions, in a context of structural racism. Episodes of micro-racism and processes of internal bordering affect black farmworkers who are finally accessing the regular rental market after years of ghettoisation in informal settlements, as explained by one of them, Zaid:

[5] Actually, migrants hosted at the Casa della Dignità can use its address for residence purposes. However, this not always applies in practice, as they are supposed to leave the Casa after six months; furthermore, the characteristics of the facility do not allow family reunification.

[6] For instance, products whose quality is good but do not comply with supermarket standards for aesthetic reasons or other reasons.

> At first, I tried to spend time in the main square in the afternoon [as an opportunity for socialisation in the village], but then I stopped. People do not like me being there. [...] You cannot see it, but I can. For example, there is a man who often smokes a cigarette at his door. Well, whenever he sees me, if I am passing by, he throws the cigarette away and leaves. Even if the cigarette is just lit ... You understand? He makes a disgusted face, spits on the floor, throws away the cigarette and goes back inside his house. There is also another man who takes his dog out for the evening walk ... he always lets the dog shit in front of my door.

Another problem that Zaid is facing, now that he has secured a stable rental contract in standard housing after years of moving from one agricultural district to another, from one ghetto to another, is that his new condition of stable residence in the village does not correspond with stable work. As he explains, he pays the rent every month, but he does not have a job throughout the whole year. Local landlords and agricultural employers, for their part, have at their disposal a pool of *stable* tenants who are also *flexible*, cheap labourers, at the same time that some local residents keep them "in check" through discourses and practices of micro-racism that ultimately justify their labour subordination.

For-profit Intermediation

Besides Casa della Dignità and the network of private housing made available through Alessandro's intermediation in San Lazzaro, farmworkers evicted from La Nuova Italia Industriale were also able to access housing in the village of San Giulio, just next to the evicted ghetto. They did it within a specific private market characterised by large grey areas. In practice, after their eviction, some of the former dwellers of La Nuova Italia have found access to (substandard) housing through regular contracts but also non-regular ones and/or through the intermediation of the same *caporali* of La Nuova Italia.

In San Giulio, a seaside village in the middle of the agricultural district of La Grande Piana, the major economic sectors are tourism and agro-industry. Importantly, the tourism-oriented character of the village is declining as a result of recent building speculation, resulting in environmental damage, which makes the seaside less attractive for tourists. In this small village where restaurants, hotels, and shops are closing, in recent years, a few new minimarkets have opened, managed by shopkeepers from

Libya, Burkina Faso, and Gambia. The public secret (which everybody knows, but nobody openly denounces) is that these new shop owners are the same *caporali* who were providing their services of intermediation to the ghetto's dwellers, who have inaugurated their legal businesses in San Giulio just after the eviction of the ghetto.

In San Giulio, new arrival infrastructures are visible in the informal use of public and private spaces. One of the African shops, which is located in a former restaurant, has turned the restaurant terrace into a space for worship where Muslim worshipers gather (most of them, migrant farmworkers), especially on Fridays. An Algerian man sporadically takes over the space in front of one of the new African shops for informal commerce of second-hand shoes, clothes, phones, and costume jewellery. Drying racks for clothes and bicycles are positioned in front of private garages, revealing the residential use of these spaces originally designed to shed cars—certainly not to shed people. The internet café of the village mostly provides administrative support to foreigners for renewal of residence permits, tax returns, and subsidy requests, among others—while "Italian" customers tend to avoid it. Finally, small groups of migrants occupy abandoned industrial buildings surrounding San Giulio, as well as uncompleted and abandoned residential projects for tourists.

Just like in San Lazzaro, one of the main vernacular accusations against immigrants in San Giulio is that their presence (rather than the environmental damage provoked by real estate speculation) lowers the value of real estate properties. At the same time, some local landlords manage to maximise the residual value of their real estate properties, even in a context of detouristification and depopulation, precisely due to migrant farmworkers' housing demand. This way, private garages are rented in response to farmworkers' housing needs; regular apartments are also rented, often without any rental contract to avoid the payment of corresponding taxes. Two of the *caporali* who are now shopkeepers also manage a few apartments, renting them to fellow countrymen farmworkers. Because turnover is frequent (due to the seasonal character of agricultural work), but migrant farmworkers do need a stable address for administrative purposes (especially to renew their residence permit), some apartments, designed to accommodate four or five tenants, are rented to nine or ten people, with landlords requiring the payment of a hundred euros per person per month—even though some tenants do not use the apartment for a large part of the year—thus extracting profit from real estate much above its market value.

Conclusions

It is now time to return to the original question at the heart of this chapter: how do migrant farmworkers access housing in a context where labour/housing intermediation is criminalised, and the emergence of new informal settlements is precluded? The case study of La Grande Piana reveals two possible - and possibly coexisting- forms in which new intermediaries emerge and old ones are still active. One is characterised by a form of "non-profit" intermediation (I place "non-profit" in quotation marks as research shows that micro-forms of profit exist also in this context): housing access is managed by the humanitarian sector, the third sector and the Church, either in guesthouses such as Casa della Dignità, or in regular houses. The other, by contrast, is eminently for profit: housing is accessed through "grey areas" of the housing market (habitational use of premises not intended for residential use such as garages, undeclared rent, overcrowded apartments), allowing local residents to maximise the profit extracted by migrant farmworkers' housing needs.

By comparing these new living conditions with the previous ones at La Nuova Italia Industriale, transformation and continuities can be identified.[7] By evicting the ghetto and preventing its reoccupation, local authorities have (largely unintentionally) initiated a process in which multiple actors can profit from migrant farmworkers' housing needs: charity organisations, individuals inspired by an humanitarian/utilitarian logic such Alessandro, and individual landlords—the latter including the same *caporali* acting as intermediaries at La Nuova Italia Industriale. For the dwellers of La Nuova Italia Industriale, leaving their shacks behind has translated into one of the following options: (1) access to a normal flat with a regular rental contract at market price (that is to say, relatively cheap); (2) access to "guesthouses" managed by the Church, in return for a relatively small fee; (3) access to the rental of substandard housing, such as garages without a kitchen and a toilet; or (4) access to a normal flat *without* regular contract, sometimes at a price that is much above the market price. While the situation of some of them has improved, for others, it has been a simple shift from substandard housing in a ghetto to substandard housing in a garage, in which they find almost equivalent conditions (e.g., no water).

[7] Importantly, it has been possible to identify continuities and transformations from farmworkers' perspective only for those farmworkers *who have remained* in the area—after the eviction, many dwellers of La Nuova Italia Industriale have indeed moved from the Grande Piana to other Italian regions, and new migrant workers have arrived.

By writing this chapter, my aim is *not* to denounce individual forms of deviancy or non-compliance with the law—as my choice to anonymise participants and localities should have made clear. By contrast, I aim to point out what happens when the phenomenon of farmworkers' exploitation and exclusion from standard housing is not identified as a *structural* phenomenon but rather described as an exceptional deviancy from the norm and attributed to individual "villains" such as *caporali*, and compliance with the law is seen as a sufficient condition to overcome structural inequalities.

The first consequence of such an oversimplified approach is that, in the absence of structural changes, the model of *caporalato* risks being reproduced—also by those who fight *against* it. New arrival infrastructures appear, in which both old and new intermediaries (former *caporali* and local residents) extract surplus value from reproductive needs of migrant workers. Similarly to what happens in migrants' smuggling, it is important to underline that the stigma attached to the figure of *caporale* by farmworkers is not necessarily and not always as negative as in policy and media discourses—also because working under full compliance with the law is not as remunerative, and can be experienced as more exploitative than *caporalato* (as Amadou's word testify). While the informal request for a fee is a necessary condition to define illegal intermediation from a legal perspective, it does not necessarily mean that all intermediaries are unscrupulous criminals. On the contrary, many other factors have to be considered, such as the amount of the surplus extracted, their professional ethics, and the legitimacy attached to them by farmworkers.

The second consequence is that a risk exists of reproducing the utilitarian logic of the (neo)liberal migration management when attempting to counter the repressive approach—as the case of Alessandro (and, overall, No CAP) testifies. Overall, migration-management rationality and the repressive/securitarian one play a prominent role in shaping the available ideological repertoire for social actors on the ground, as it concerns their representations of migration and their intervention on it. Even the most progressive political leaders, members of radical social movements committed to social justice issues, and representatives of civil-society organisations (including migrants' organisations), in their effort to counter restrictive and securitarian migration policies, might unwittingly reproduce the utilitarian logic underlying the migration-management discourse—as the "win-win" rationale in Alessandro's discourse (and praxis) shows. While social theory has long shown that a component of opportunism exists in all social relations, the utilitarian logic in migration management becomes particularly problematic when it is presented as a form of justice-oriented, socially transformative approach (such as the "ethical label" for supermarket products), and when it arises from a misrepresentation of farmworkers as victims without agency, at times even silencing their voice. This was the case in the exclusion of migrant farmworkers' voices in the anti-eviction protests at La Nuova Italia Industriale, and in the establishment of the "fair price" for "ethical" products.

Acknowledgments The author gives thanks to Domenico Claudio Perrotta for his help to access the field and his insights on the reproduction of the *caporalato* system shared in private conversations, and to the editors Natalia Buier, Theodora Vetta, and Susana Narotzky for their valuable support throughout the writing and editorial process.

References

Altenried, M., Bojadžijev, M., Höfler, L., Mezzadra, S., & Wallis, M. (2018). Logistical borderscapes: Politics and mediation of mobile labor in Germany after the "summer of migration". *South Atlantic Quarterly, 117*(2), 291–312.

Balibar, É., & Wallerstein, I. (1988). *Race, Nation, Classe: Les Identités Ambiguës.* La Découverte.

Caruso, F. S., & Corrado, A. (2021). *Essenziali Ma Invisibili. Analisi Delle Politiche e Delle Iniziative Di Contrasto Allo Sfruttamento e per l'inclusione Di Lavoratori Migranti Nel Sud Italia.* Rosenberg & Sellier.

Caruso, F. S., & Macciani, C. (2023). The Criminalizatio of labor intermediation in Southern Italy countryside: An ethnographic account of the evolution of rural informal settlements and Caporalato in Foggia. *Studi sulla questione criminale, 18*(3), 121–142.

Corrado, A., et al. (2018). *Migrazioni e Lavoro Agricolo in Italia: Le Ragioni Di Una Relazione Problematica.* Open Society Foundations and European Policy Institute. Available at https://www.opensocietyfoundations.org/uploads/b6ead456-1d7e-443e-b066-8f0e70a6f1be/is-italian-agriculture-a-pull-factor-for-irregular-migration-report-it-20181205.pdf

De Genova, N. (2013). The perplexities of mobility. In O. Söderström et al. (Eds.), *Critical mobilities* (pp. 101–122). Routledge.

Dines, N. (2018). Humanitarian Reason and the Representation and Management of Migrant Agricultural Labour. THEOMAI Journal Estudios críticos sobre Sociedad y Desarrollo / Critical Studies about Society and Development, 38. http://www.revista-theomai.unq.edu.ar/numero38

Dines, N. (2023). After entry: Humanitarian exploitation and migrant labour in the fields of Southern Italy. *Environment and Planning D: Society and Space, 41*(1), 74–91.

Dines, N., & Rigo, E. (2015). Postcolonial Citizenships and the 'Refugeeization' of the Workforce: Migrant Agricultural Labor in the Italian Mezzogiorno. In S. Ponzanesi & G. Colpani (Eds.), *Postcolonial Transitions in Europe: Contexts, Practices and Politics* (pp. 151–172). Lanham, MD: Rowman and International.

Fanizza, F., & Omizzolo, M. (2018). *Caporalato: An authentic Agromafia*. Mimesis International.

Fassin, D. (2012). *Humanitarian reason. A moral history of the present*. University of California Press.

GCIM. (2005). *Migration in an interconnected world – New directions for action: Report of the Global Commission on International Migration*.

Geiger, M., & Pécoud, A. (Eds.). (2010). *The politics of international migration management*. Palgrave Macmillan.

Grappi, G. (2020). Europe and the Contested Politics of migration: Between logistification and global justice. *SSRN Electronic Journal* (March).

Kalir, B. (2019). Departheid: The Draconian governance of illegalized migrants in Western states. *Conflict and Society, 5*(1), 19–40.

Kalm, S. (2010). Liberalizing movements? The political rationality of global migration management. In M. Geiger & A. Pécoud (Eds.), *The politics of international migration management* (pp. 21–44). Palgrave Macmillan.

Maaroufi, M. (2017). Precarious integration: Labour market policies for refugees or refugee policies for the German labour market. *Refugee Review, 3*, 15–33.

McMichael, P. (2009). A Food Regime Genealogy, *In Journal of Peasant Studies, 36*(1), 139-170.

Meeus, B., Arnaut, K., & van Heur, B. (Eds.). (2019). *Arrival infrastructures: Migration and urban social mobilities*. Palgrave Macmillan.

Mwambari, D. (2023). *Navigating cultural memory: Commemoration and narrative in postgenocide Rwanda*. Oxford University Press, Incorporated. http://ebookcentral.proquest.com/lib/kuleuvenul/detail.action?docID=7256074

Papadopoulos, A. G., & Fratsea, L.-M. (2017). Migrant labour and intensive agricultural production in Greece: The case of Manolada strawberry industry. In A. Corrado, D. Perrotta, & C. De Castro (Eds.), *Migration and agriculture: Mobility and change in the Mediterranean area* (pp. 128–144). Routledge.

Perrotta, D. (2014). Vecchi e Nuovi Mediatori. Storia, Geografia Ed Etnografia Del Caporalato in Agricoltura. *Meridiana, 79*, 193–220.

Perrotta, D. (2016). Processing tomatoes in the era of the retailing revolution. Mechanization and migrant labour in Northern and Southern Italy. In

A. Corrado, C. De Castro, & D. Perrotta (Eds.), *Migration and agriculture. Mobility and change in the Mediterranean area* (pp. 58–75). Routledge.

Rigo, E., & Dines, N. (2016). Campi per 'Cittadini Migranti Stagionali.' *Connessioni Precarie.*

Rose, N., & Miller, P. (2010). Political power beyond the state: Problematics of government. *The British Journal of Sociology, 61*(s1), 271–303.

Schneider, H. (2022). Local migration regimes in rural areas: The example of refugee reception in Saxony, Germany. *Comparative Population Studies, 47*, 413–442.

Wessendorf, S., & Phillimore, J. (2019). New migrants' social integration, embedding and emplacement in superdiverse contexts. *Sociology, 53*(1), 123–138.

Xiang, B., & Lindquist, J. (2014). Migration infrastructure. *International Migration Review, 48*(1_suppl), 122–148. https://doi.org/10.1111/imre.12141

Appropriated, Segregated, Hierarchised: The Spaces of Migrantised Agroindustry in Mediterranean Europe

Juan Castillo-Rojas-Marcos, Yoan Molinero-Gerbeau, and Gennaro Avallone

Introduction

The current European division of labour positions the Mediterranean countries, notably Spain and Italy, and, to a much lesser extent but with a similar structure, Greece, as major suppliers of fresh fruit and vegetables to the single market (Fritsch & Matthes, 2020). Benefitting from an optimal climate and favourable socio-political conditions, such as access to EU Common Agricultural Policy (CAP) funds, these countries have witnessed

J. Castillo-Rojas-Marcos (✉) • Y. Molinero-Gerbeau
Universidad Pontificia Comillas, IUEM (Instituto Universitario de Estudios sobre Migraciones), Madrid, Spain
e-mail: jcastillo@comillas.edu; ymolinero@comillas.edu

G. Avallone
Department of Political and Social Studies, University of Salerno, Salerno, Italy
e-mail: gavallone@unisa.it

S. Narotzky et al. (eds.), *Agricultural Extractivism in the Mediterranean Region*,
https://doi.org/10.1007/978-3-032-05599-6_7

the proliferation of globalised agroindustrial enclaves, promoting the mass production of food for the European market (Moraes et al., 2012).

This agroindustrial development led to substantial social and productive structural transformations in the rural areas of Mediterranean Europe. Compared to previous peasant forms, this model intensifies the double appropriation of nature (i.e., the exploitation of labour and the predation of non-human nature) through monocultures, agrochemical inputs, Fordist rationalisation of work processes, and migrant proletarianisation (Avallone, 2014; Rodríguez-Calles, 2022). This socio-environmental extractivism turns both bodies and territories into expendable resources at the disposal of capital accumulation (Wallerstein, 2010; De Castro, 2014; Moore, 2015).

Every society develops a specific way of organising the spaces it inhabits (Lefebvre, 1974). While the social production of space is complex, multicausal, and shaped by diverse (including subaltern) actors, one structural factor remains consistently central: shifts in the mode of production or social formation (Smith, 1984). This raises the question of how rural spaces in Mediterranean Europe have been reshaped by the transition to an industrialised, waged, export-driven, migrant labour-based agriculture.

One of the most significant transformations driven by agroindustry has been the rapid incorporation of large numbers of migrant labourers into its workforce—and, by extension, into the demographic makeup of Euro-Mediterranean rural regions (Nori & Farinella, 2020; Papadopoulos & Fratsea, 2024; Papadopoulos et al., 2021; Moraes et al., 2012; Molinero-Gerbeau, 2020). Numerous studies examine these workers' experiences, often focusing on legal incorporation mechanisms, labour exploitation, and health-related risks (Sajir et al., 2022; Escrivà, 2022).

Other studies have examined workers' residential spaces, often focusing on their substandard conditions and/or connecting them to the main themes mentioned above (Caruso, 2018; Papadopoulos & Fratsea, 2021). Despite existing research, there remains a lack of comprehensive analyses addressing the spatial reconfiguration in territories shaped by agro-enclaves, the specific residential settings occupied by migrant workers within these areas, the intersection of these spatial arrangements with the forms of exclusion they experience, and the spatial production processes that give rise to these dynamics.

This chapter contributes to a theoretical understanding of the mode of appropriation underpinning capital accumulation in Euro-Mediterranean agroindustry, focusing on its spatial dimension. It argues that this model entails a triple appropriation: of human nature, non-human nature, and of social space: not just the territory and natural environment but also buildings, infrastructure, public areas, spatial demarcations (material or symbolic), the space-shaping practices of their inhabitants, and the systemic relationships among all these elements (Lefebvre, 1974).

To do so, the chapter starts with a theoretical analysis discussing the global capitalist crisis of appropriation, arguing that agroindustry is a form of agrarian extractivism that emerged in response to that crisis. It then examines the development of the agroindustrial model in Mediterranean Europe, highlighting its technological, labour, ecological, and social dimensions, and reflects on the forms of spatiality generated by comparable extractive industries in other contexts.

These discussions lay the groundwork for an empirical analysis of how social space has been transformed in agroindustrial enclaves in Spain, Italy, and Greece, based on a systematic literature review using Scopus and Web of Science. The search employed terms from the semantic fields of "agroindustry"[1] and "migrant worker",[2] restricted to the period between 2018 and 2024. Texts lacking specific reference to sociospatial issues were subsequently excluded, yielding a final sample of 63 publications.[3] A thematic analysis identified spatial restructuring patterns, complemented by additional statistics and peer-reviewed sources when needed to triangulate interpretations and strengthen the arguments.

Agroindustry in the Global Crisis of Appropriation

Global capitalism has historically rested on a socio-ecological relation based on the cheap appropriation of human and extra-human natures. Economic valorisation is grounded in the appropriation of unpaid labour, human and non-human, including the categories identified by Mies (1986): women, nature, and colonies. The essential basis of life production through unpaid labour enables the exploitation of paid labour and sustains capital accumulation (Moore, 2015).

[1] Those are: agroindustry; agriculture; "primary sector"; agrifood; horticulture; "food production"; agribusiness.

[2] Those are: "migrant work"; "migrant labour"; "seasonal workers"; "circular migration"; "temporary workers"; migration; foreigners; "foreign workers"; migrants.

[3] 25 about Spain, 22 about Italy, 9 about Greece, and 9 about two or more of these countries.

This process entails a symbolic and political devaluation of nature, a reduction that allows its appropriation at minimal cost. What is valued—capital—is separated from what is cheapened—nature—thus guaranteeing accumulation rates necessary for the system's reproduction, as a rate of 3% accumulation, as Harvey notes, is considered essential for capitalism (Harvey, 2010). Such accumulation has historically been secured through continually shifting frontiers of appropriation, seeking new spaces and forms of cheap labour and energy. These frontiers include not only new geographical areas but also internal divisions within states, such as the legal and symbolic marginalisation of migrant populations (Sayad, 1984).

However, world-ecology analysis suggests that the current era, marked by climate change, signals the end of the epoch of cheap natures. The declining ecological surplus—evidenced by the rising cost of labour, food, energy, and raw materials—is transforming the web of life from a source of capital accumulation into its obstacle (Moore, 2017). This "biotarian revolt", or the refusal or inability of historically cheap natures to serve capital, finds expression in ecological degradation, resource depletion, and social resistance, including labour and social movements (Moore, 2022).

This blockage triggers, in turn, new capitalist strategies for pushing frontiers and trying to reinvigorate capital accumulation. Agriculture exemplifies this crisis of appropriation and capital's rearticulations in response to it. The exhaustion of the agroecological frontiers first activated during the long sixteenth century has led to efforts to expand the frontier of technical control, degrading and deskilling labour in the process, displacing skilled with semi-skilled labour in favour of more vulnerable and exploitable working populations, and reinforcing extractivist dynamics in food systems (Moore, 2010).

Factory-like production methods were already applied to agriculture in California in the early twentieth century, but their global diffusion accelerated in the 1970s with the so-called Green Revolution (Zapata Rivera, 2022). The agroindustrial model emerged in this context, marking a transformation of agriculture characterised by increased labour productivity, purchased inputs, mechanisation, land reorganisation, international markets, and the emergence of large corporations (FitzSimmons, 1986).

A defining feature of the agroindustrial model is its structural dependence on wage labour, turning farms into "food factories" and fostering a new international rural proletariat (Avallone, 2014). With input costs fixed by suppliers and retail prices set by supermarket chains, labour is the main adjustable cost. Profit and accumulation thus depend on reducing labour costs, often by exploiting precarious or undocumented migrants. This creates a labour market where the "labour norm" (De Castro, 2014) is shaped by local business practices rather than collective agreements. State inaction or complicity enables irregular employment, allowing underpayment, lack of social contributions, and excessive working hours (Molinero Gerbeau & Avallone, 2016). Even regularised workers are subject to "grey labour" practices (such as undeclared hours or underreported workdays) that reduce wages and social contributions (Avallone, 2017).

Those technologies, production schemes, and social relations define the mode of appropriation of agroindustry. Gudynas (2016) describes modes of appropriation as the specific way an industry organises the extraction of nature under its domain. This includes not only productive relations in the traditional sense but also the full range of operations and mechanisms—institutional, ecological, and symbolic—that underpin its extractive activity (Gudynas, 2016).

While extractivist modes of appropriation have shaped capitalist agricultural practices since the sixteenth century (Acosta, 2013), in recent decades, the agroindustrial model has intensified and reshaped these dynamics. Some of its features, like export orientation, enclave production, intensive monocultures, reliance on machinery and agrochemicals that degrade soils and ecosystems, depeasantisation, and worsening labour conditions define contemporary agrarian extractivism (Gudynas, 2015; McKay & Colque, 2022). The next section explores its development in Spain, Italy, and Greece.

The Development of Agroindustry in the Euro-Mediterranean Area

The expansion of agroindustry in Mediterranean Europe has been, in different ways, state-supported and aligned with EU policy frameworks geared towards maximising agricultural output at minimal cost. This transformation took place against a backdrop of rural exodus in Europe (Newsham & Rowe, 2023), paradoxically coinciding with increased demand for agricultural labour. Traditional internal labour circuits, previously sufficient for meeting the needs of mostly family-run farms, began to fail: demand grew increasingly exceeding capacity and, at the same time, part of the native agricultural workforce was shifting to better-valued employment sectors. International migration emerged as the primary mechanism to resolve this contradiction. Since the 1990s, migrant labour has become structurally embedded in agricultural production in Greece, Spain, and Italy, becoming the backbone of the sector's wage labour force (Nori & Farinella, 2020; Papadopoulos & Fratsea, 2024; Papadopoulos et al., 2021; Moraes et al., 2012; Molinero-Gerbeau, 2020).

Agroindustrial enclaves in Mediterranean Europe became globalised not only through international agribusiness and export infrastructures but also via the increasing migrantisation of their workforces (Molinero-Gerbeau, 2020). This was achieved mainly through irregular labour migration in Italy and Greece, while regular, institutionalised channels played a bigger role in Spain (Corrado & Caruso, 2022; Kasimis et al., 2015). Simultaneously, environmental degradation intensified as a result of the overexploitation of land and water, and of pollution from the depletion of nitrates and other agrochemicals (Rodríguez-Calles, 2022; Martínez-Valderrama et al., 2023).

The economic viability of agroindustry rests on a complex matrix of scientific, technical, and institutional supports. These include high technification, input dependency, European subsidies under the Common Agricultural Policy (CAP), and various national mechanisms (Bigiani et al., 2023). Nevertheless, the model's profitability is fundamentally rooted in large-scale production with the lowest possible labour costs. In Italy and Greece, labour costs are kept low by the widespread occurrence of "informal-to-grey" labour arrangements, often combined with the irregular administrative situation of the workers themselves (Avallone, 2017; Kasimis et al., 2015).

In Spain, those situations coexist with a higher incidence of more formalised labour relations. Nonetheless, this does not necessarily mean less exploitation and structural violence, as illustrated by the GECCO mechanism, under which seasonal workers can be recruited for up to nine months, brought in during harvest periods, and repatriated afterwards. This scheme institutionalises extreme dependency, as each worker's legal stay in Spain depends on their employers' approval, resulting in frequent abuses like unpaid overtime, contract violations, and even sexual harassment (Castillero, 2021; Reigada, 2022).

This Mediterranean agroindustrial model has significant, harmful impacts on the lives and bodies of workers, especially migrants and, even more so, those who are vulnerable in terms of gender, "race", and/or administrative status. First, socio-economic conditions are marked by structural job instability, low income, and employer control mechanisms that reinforce subordination. The intersection of high dependence on the employer and administrative precariousness leads to situations of extreme disempowerment of migrant proletarians, as illustrated by the caporalato system in southern Italy or the precarious permits granted to labourers on a temporary basis in agricultural enclaves in Greece (Melossi, 2021; Gialis & Herod, 2014).

Second, intensive agricultural work has a direct impact on the physical and mental health of workers, who are exposed to long working hours, extreme temperatures, toxic substances, and extremely limited medical care. The most common pathologies include muscle pain, untreated injuries, exhaustion, stress, psychosomatic symptoms, and reactions to toxics (Kukreja, 2021a; Escrivà, 2022). Structural barriers to accessing health systems and social services, which exclude high percentages of migrant workers (up to 60% in the case of Italy, according to a 2018 study) aggravate these conditions by preventing access to the necessary care for a full and timely recovery (Ceccarelli & Ciconte, 2018; Kotsila & Kallis, 2019).

Migrant women also face specific forms of violence that exacerbate their vulnerability. This is particularly evident in the agroindustrial enclave of Huelva (Spain), where the unusually high proportion of women in the workforce has been accompanied by numerous cases of sexual violence and coercion against Moroccan women workers, as well as the application of sexist criteria in their recruitment (Castillero, 2021; Reigada, 2022).

Extractivism and Production of Space

The previous section outlined key elements of Mediterranean agroindustry's mode of appropriation: its technological and productive schemes, its institutional framework, ecological damage, and the migrant-based, oppressive labour regime. However, one crucial aspect remains untackled: spatialisation, or how this model reshapes the spaces it inhabits. Here, space is not merely the surface on which society settles, nor its passive backdrop, but the environment with which society interacts, on which it depends and of which it is a fundamental component: the social space (Lefebvre, 1974).

Social space includes a multiplicity of interconnected layers: ecosystems (modified by human societies), built environments, infrastructure, symbolic and functional spatial divisions, and the human movements and practices that shape space. The artificial nature of these elements, and the fact that in every geo-historical context they are uniquely designed and structured, reveals that each society—or, more precisely, the contending social classes and groups within it—produces the space it inhabits (Lefebvre, 1970, 1974).

The social production of space typically unfolds through intricate, multicausal dynamics involving a broad spectrum of actors, including subaltern groups; yet, a structural determinant invariably exerts a central influence: changes in the mode of production and/or social formation (Smith, 1984; Lefebvre, 1970). The notion of social formation refers to the observable and concrete social, economic, and political system in which a mode of production and social reproduction is incarnated in a given historical and geographical context (Harnecker, 1971). And, as has been argued, a defining feature of the agroindustrial model in abstract terms and the Euro-Mediterranean agroindustrial social formation in particular is their extractivist mode of appropriation.

Scientific literature on extractive industries shows they typically expand within their operating space, transforming not just production areas but reconfiguring the territory as a whole in various manners. First, they tend to drastically alter the surrounding landscapes, both rural and urban, either to better adapt them to their productive needs or due to their unintended but unchecked effects, or both (Svampa, 2013; Gudynas, 2010). Second, such transformations generate new sources of inequality and exclusion, as the new spatial arrangements can limit the opportunities and disrupt the social reproduction activities of local disadvantaged groups, often along "racial", (post)colonial lines (Bebbington & Bury, 2013; McKay & Colque, 2022). And third, new geographies of disease and death proliferate, as the industries' pollution and impact on health concentrate in areas inhabited by those same populations (Murrey & Mollet, 2023; Bebbington & Bury, 2013).

In summary, extractivism tends to generate sociospatial transformations that reduce the inhabitability of spaces, that is, their capacity to function as habitats. In critical architecture and urbanism, habitats are conceptualised as the expanded living space (not just houses but also the whole set of public and semi-public spaces where daily life takes place) of individuals and communities that adequately sustains their material, cultural, and relational needs; their health; and their social reproduction activities (Echeverría Ramírez, 2003; Echeverría Ramírez et al., 2009). The next section will analyse how and to what extent this pattern of decreased inhabitability emerges in the case of the agroindustry of Mediterranean Europe, based on the scientific evidence available.

Mediterranean Agroindustry as a Space Producer

The reorganisation of the structure of land property, particularly in the form of land concentration and land enclosure or privatisation (interrelated forms of land dispossession), is a significant element of the development of agroindustry (FitzSimmons, 1986; McMichael, 2021). Although the literature reviewed does not generally give priority attention to this issue, there are a considerable number of occasional or indirect references suggesting that this spatial dynamic has also been present in Mediterranean Europe. In terms of land concentration, Raeymaekers (2024) mentions that the average size of agricultural operations in Italy has grown from 5.3 hectares in 1990 to 11.1 hectares in 2020. Similar tendencies of land concentration are mentioned regarding Greece (Papadopoulos et al., 2021); the Spanish enclaves of Lleida (Kotsila & Argüelles, 2023), Huelva (Reigada, 2022), and Spain as a whole (Garcés & Güell, 2021); and the Saluzzo region (Brovia & Piro, 2021), the Lazio (Salvia, 2020), and Italy as a whole (Melossi, 2021).[4]

In other words, the share of the land held by small farmers has shrunk across Mediterranean Europe, while large property has undergone a gradual expansion. In the agroindustrial model, competition for global export markets is intense, and small family farms face structural disadvantages compared to large corporate holdings (Mozo et al., 2022). As a result, the latter increasingly capture enclave land, deepening local inequalities and undermining the material foundations of more sustainable, small-scale food channels. This sociospatial shift away from micro-property and locally oriented rural economies is ongoing across the region, with more advanced developments in Spain and Italy than in Greece (Papadopoulos et al., 2021; Sajir et al., 2022).

Land enclosure and privatisation appear less central to Euro-Mediterranean agroindustry, as they are rarely discussed in the reviewed literature. This is unsurprising, given that such dispossession is usually linked to regions like ex-socialist Central and Eastern Europe (Bański, 2017) or former colonial peripheries with vibrant communal economies (Chung, 2017; Baka, 2017). Yet, Huelva (Spain) stands out as an exception. Reigada (2022) notes that much of today's agroenclave expands over land that until the 1980s was public or communal. At that time, the recent

[4] Nonetheless, in most cases those tendencies are mentioned briefly and superficially, just supporting arguments about other issues.

introduction of agroindustrial production methods into the area (such as greenhouses techniques, irrigation systems, and agrochemicals) exponentially increased the potential profitability of those common lands. These lands were subsequently appropriated (first informally, later officially state-sanctioned) by locals who, through this process, became the founders of the present-day agroindustrial capitalist class, in Huelva enclave's particular episode of primitive accumulation (Reigada, 2022, pp. 46–58).

Although the literature doesn't document similar processes in other enclaves, agrarian statistics from Italy and Spain suggest they may be more widespread.[5] In both countries, the share of public and common land has declined over decades (Istat, 2010, 2020; INE, 1999, 2009),[6] and major agroindustrial enclaves often show significantly lower shares than national averages (Istat, 2020; INE, 2020).[7] Empirical research assessing whether these trends reflect land privatisation linked to agroindustrial enclave development would offer a crucial, potentially ground-breaking contribution to the literature.

Another key issue is the spatial allocation of the migrant proletariat within agroindustrial enclaves. Studies identify a widespread housing model for migrant farmworkers across Europe: employer-provided accommodation on farms (Ruíz Ramírez et al., 2024). Typically, these are small barracks or prefabricated metal or plastic units with bunkbeds. Such housing is often overcrowded and lacks basic necessities like electricity, drinking water, or cooking equipment (Kukreja, 2021b; Molinero-Gerbeau, 2021; Brovia & Piro, 2021). Even when those material deficiencies are avoided, two issues persist. One is social isolation, as facilities are in remote rural areas, segregated from locals and lacking services or public spaces. The other is increased dependency, as employers also act as landlords, gatekeepers, and often the sole transport link to town (Kukreja, 2021b; Reigada, 2022).

[5] No equivalent data could be found regarding Greece.

[6] The information is fragmentary, because agricultural censuses have not always asked (and/or made public the data regarding) the form of ownership (public/common or private) of farms. Based on the cases with information available, in Spain, the percentage of land under non-private forms of ownership fell from 25.2% in 1999 to 14% in 2009 (INE, 1999, 2009), and in Italy, it fell from 12.7% in 2010 to 8.9% in 2020 (Istat, 2010, 2020).

[7] In Spain, with a national average of 14%, Huelva has a 5.7% and Murcia a 2.45% (INE, 2009); in Italy, with a national average of 8.9%, Ragusa has a 0.1% and Mantova a 2.8% (Istat, 2020).

Specifically in Mediterranean Europe, another frequent model is the rural slum: clusters of makeshift dwellings built from plastic, cardboard, and pallets, marked by extreme material precarity (Fravega, 2023; Kotsila & Kallis, 2019; Rubio González et al., 2023). These areas share—and worsen—the deficiencies of employer-provided housing: overcrowding, remoteness, and lack of infrastructure. However, being self-managed and free from employer control, they permit greater social interaction and autonomy, posing less structural barriers to the emergence of social solidarity (Bonnin et al., 2024).

Less common residential models include public collective shelters, as seen in Lleida and Jaén, Spain, and Apulia, Italy (Garcés & Güell, 2021; Caruso, 2018), and state-run, refugee camp-like tent settlements, like in Saluzzo, Italy, and Myrsini, Greece (Brovia & Piro, 2021; Papadopoulos & Fratsea, 2024). These resemble, respectively, on-farm facilities and rural slums, differing in their state-managed nature. Despite this, they follow the same pattern of remoteness and segregation, being located outside towns or in peripheral areas. Additionally, in Murcia, Spain (Rodríguez-Calles & Iglesias, 2025; Rodríguez-Calles, 2023) and different Greek sites (Kotsila & Kallis, 2019; Papadopoulos & Fratsea, 2021, 2024), migrant farmworkers are reported living in towns. However, they are typically concentrated in deprived outskirts or small, remote hamlets, often in overcrowded and/or semi-abandoned rural housing. Though remoteness may be reduced, segregation persists.

Another defining trait of migrant farmworker spaces is the structural proliferation of sources of infection and danger. Material precarity and overcrowding in rural slums and on-farm facilities, combined with selective state neglect, allow preventable, life-threatening conditions to persist. Many areas, inhabited for years, are deprived of waste-collection and street-cleaning services, creating extreme unsanitary environments (Fravega, 2023; Molinero-Gerbeau et al., 2021). Waste accumulates near dwellings, while toilets and baths are often improvised, communal, outdoors, and adjacent to living quarters (Kukreja, 2021a; Macciani, 2024). These arrangements attract pests, facilitate the spread of infections, and cause water stagnation and waste, further hindering hygiene. Combined with intensive agrochemical use contaminating land, water, and bodies, this creates severe soil and health degradation. New death and illness geographies emerge, as these hazardous conditions are mainly concentrated in areas where migrant rural proletarians live and work - spaces , largely invisible to the native population.

This unequal, ethnically/"racially" hierarchised spatial structure has tangible health impacts, as shown by much of the reviewed literature. In Spanish agroenclaves, Rodríguez-Guerrero et al. (2024) reported skin diseases from polluted water; in Manolada (Greece), Kukreja (2021b) noted frequent fevers and diarrhoea; and in the Evrotas Delta, Kotsila and Kallis (2019) documented repeated malaria outbreaks over five years. The COVID-19 pandemic further victimised already exposed migrant farmworkers, who in some cases faced higher infection rates than the general population (Kukreja, 2021a; Corrado & Caruso, 2022; Pedreño Cánovas et al., 2022). While all farmworkers, including natives, endured hazardous conditions (on-site work, employers ignoring safety measures), migrants were sociospatially further exposed: overcrowded housing, limited healthcare access, and the impossibility of confinement in homes without water, basic amenities, or nearby stores.

One particularly frequent and life-threatening risk is large-scale fires. Especially (though not exclusively) in rural slums, dwellings periodically ignite, with flames rapidly consuming homes, belongings, and lives. While immediate causes vary, the underlying conditions are structurally constant: hundreds or thousands living outdoors in flammable, precarious shelters, under the hot, dry climate of Mediterranean Europe (Dines, 2023; Kukreja, 2024; Garcés & Güell, 2021).

Various sources of pollution, disease, and deprivation converge to undermine the habitability of territories that host agroindustrial enclaves—an uninhabitability itself subject to stratification along intersecting lines of class and "race"/ethnicity. "Racial" categorisation plays a distinct role: evidence suggests that the forms and intensity of violence and segregation faced by migrant farmworkers vary by perceived whiteness. Non-white workers face specific discrimination in the housing market, further limiting their chances of escaping confinement in remote, isolated settings like rual slums or on-farm facilities (Ambrosini, 2024; Castillero, 2021; Kotsila & Kallis, 2019).

Consistently, several texts associate informal settlements and makeshift shacks with Black and Brown African and Asian workers, while White Eastern Europeans tend to avoid them (Buier, 2024; Melossi, 2021; Papadopoulos & Fratsea, 2021, 2024).[8] This alone exposes racialised

[8] It is significant that, when discussing the settlements in Huelva (Spain), Buier (2024, p. 47) explains that they are inhabited "in the majority by African men, Moroccan women, and racialized Eastern Europeans", suggesting that, among people from countries such as

groups to the most harmful forms of segregated health and socio-economic impacts. Moreover, since the onset of large-scale labour migration in agroindustry, there have been repeated episodes of pogrom-like violence. Nori and Farinella (2020) cite El Ejido (Spain, 2001), Castelvolturno (Italy, 2008), Rosarno (Italy, 2010), and Manolada (Greece, 2013). A clear pattern: in every case, victims belonged to racialised groups, not white Eastern Europeans—despite the latter's major presence within the region's migrant farm workforce.

The spatialities of agroindustry reflect geographies of non-being, structured by a "violent stratification of humanity" where the racialised Other holds a position "more vulnerable to extermination" (Rodrigues de Azevedo, 2023a). As Corrado et al. (2018, p. 24) note, there is an "ambiguous coexistence of economic demand for migrant labour in the fields and social hostility to their presence in the streets", especially towards non-white workers. Migrant (and, particularly, racialised) proletarians are relegated to invisible spaces—indispensable for their labour yet unwanted in public life—where even the most basic institutional mechanisms for sustaining life are withdrawn. The next section systematises and explores the implications of these findings.

Discussion and Conclusions

Mediterranean agroindustry's mode of appropriation involves certain patterns of multidimensional restructuring of the spaces in which it operates, closely related to the social, economic, and environmental transformations associated with this model. Land distribution (and, consequently, access to and use of land) is reconfigured. New types of residential and everyday life spaces emerge, most notably rural slums. The territory becomes fractured into areas assigned to different population categories, defined by intersecting class and "race"/ethnicity divisions. The possibility of leading healthy, secure lives is similarly stratified across these areas, and gradually diminished throughout enclaves as a whole. Taken together, these tendencies reveal Mediterranean agroindustry as a powerful space producer.

Romania and Bulgaria, Roma workers end up in the settlements, while white workers elude them. This alone reinforces the argument that it is not simply a question of citizenship but of "race", which places white workers in a structurally more favourable position to avoid the most harmful forms of segregation in Euro-Mediterranean enclaves.

These patterns are not static or fixed outcomes of past transformations resembling a "primitive accumulation" dynamic (Marx, 1867). The socio-spatial phenomena discussed are ongoing, expanding processes, persisting as long as agroindustry remains active in each territory. Mediterranean agro-enclaves depend on such sociospatial rearrangements to operate. In this sense, the industry's mode of appropriation entails a triple movement, simultaneously incorporating human nature, non-human natures, and the spaces within which they interact.

This mode of appropriation has historically succeeded in sustaining capital accumulation, and agriculture in the enclaves located within the studied states is highly profitable. Its export focus and industrial structure highlight its profit-making capacity. But does agroindustry in Mediterranean Europe inherently demand such dire living and working conditions? Is the production of segregated, racialised, uninhabitable space intrinsic to it, or could it coexist with more inclusive spatial arrangements? In short, is this ecological-economic model inherently unsustainable, socially, spatially, and ecologically?

The key lies in the power relations between labour, farmers/agricultural companies, and other value chain actors (industrial input suppliers, food distributors from Northern Europe, etc.). With limited bargaining power over the latter, farmers shift all pressure onto the former to generate profit. Class struggle thus unfolds between farms and farmworkers, requiring the devaluation of labour both politically and symbolically. In line with Jason W. Moore's analysis:

> For capitalism, Nature is "cheap" in a double sense: to make Nature's elements "cheap" in price; and also to cheapen, to degrade or to render inferior in an ethical-political sense, the better to make Nature cheap in price. These two moments are entwined at every moment and in every major capitalist transformation of the past five centuries. (Moore, 2016, p. 3)

The reduction of the political-symbolic value of the workforce occurs through various strategies and conditions, including migration policies, housing conditions, and spatial exclusion and invisibilisation. The more these conditions produce a differentiated and devalued working population, the more this population will be reduced in its overall value. Combined with other concurrent factors—most notably what Rodrigues de Azevedo (2023a, 2023b) and Fanon (1967), describe as a capilar "fear of the black man," that is, a visceral rejection of the public presence and visibility of non-white bodies, fiercely adhered to the cultural foundations of these European societies,[9] the result is the production of a segregated, unequally inhabitable space.

In sum, similarly to other extractive industries (Svampa, 2013; Gudynas, 2010), the agro-enclaves' social geography is radically altered, including not just productive sites but also surrounding areas (residential and otherwise), and the patterns of segmentation, segregation, and hierarchisation that govern the whole territory. Space itself is appropriated, rearticulated according to an extractive logic: rearranged in its uses, functions, and property distribution to better adapt to the productive necessities of agroindustry, and incorporated as a means of production actively intervening in agroindustrial valorisation processes. Additionally, in the areas associated with the migrant (and, even more, racialised) proletariat, space loses its potential to function as a habitat (Echeverría Ramírez, 2003; Echeverría Ramírez et al., 2009), due to the uncontrolled effects of the extractive activity and in line with the symbolic order of dehumanisation of these subjects, a central aspect of this social formation.

This discussion allows us to understand that, similarly to other social formations articulated around extractivism, in addition to proletarian bodies and extra-human nature, Euro-Mediterranean agroindustry also subjects places to predation. It devours whatever form of producing space societies may have articulated over historical time. Instead, it excretes territories organised on the basis of productive criteria and (post)colonial hierarchies. It devours habitats and excretes open-air assembly lines.

Competing Interests The research leading to these results received funding from the European Union's Horizon Europe Programme for research project SAFEHABITUS (Strengthening Farm Health And Safety Knowledge And Innovation Systems, under grant agreement no. 101084270).

References

Acosta, A. (2013). De cómo el extractivismo oprime el cerebro de nuestras sociedades: Algunas anotaciones al caso ecuatoriano. *Onteaiken: Boletín de Prácticas y Estudios de Acción Colectiva, 15*, 113–122.

[9] This might include the housing discrimination against non-whites in the enclaves or the systematic withdrawal of public services from migrant worker-inhabited zones.

Ambrosini, M. (2024). Immigrant work and the production of Italian Agrifood: The variants of subordinate integration. *Journal of Immigrant & Refugee Studies*, 1–16. https://doi.org/10.1080/15562948.2024.2424164

Avallone, G. (2014). Migraciones y Agricultura En Europa Del Sur: Emergencia de Un Nuevo Proletariado Internacional. *Migraciones Internacionales, 27*, 137–169. https://doi.org/10.17428/rmi.v7i27.659

Avallone, G. (2017). The land of informal intermediation. The social regulation of migrant agricultural labour in the Piana Del Sele. In A. Corrado, C. De Castro, & D. Perrotta (Eds.), *Migration and agriculture. Mobility and change in the Mediterranean area* (pp. 217–230). Routledge.

Baka, J. (2017). Making space for energy: Wasteland development, enclosures, and energy dispossessions. *Antipode, 49*(4), 977–996. https://doi.org/10.1111/anti.12219

Bański, J. (2017). The consequences of changes of ownership for agricultural land use in Central European countries following the collapse of the Eastern Bloc. *Land Use Policy, 66*, 120–130. https://doi.org/10.1016/j.landusepol.2017.04.045

Bebbington, A., & Bury, J. (Eds.). (2013). *Subterranean struggles: New dynamics of mining, oil, and gas in Latin America*. University of Texas Press.

Bigiani, L., Antonioli, F., & Severini, S. (2023). The impact of CAP subsidies on the productivity of cereal farms in six European countries: A historical perspective (2008–2018). *Food Policy, 119*, 102473. https://doi.org/10.1016/j.foodpol.2023.102473

Bonnin, I., Fravega, E., & Palmas, L. Q. (2024). Of land and sea. "Interested solidarities" and the migration industry from below. *Mondi Migranti, 2*, 159–189. https://doi.org/10.3280/mm2024-002007

Brovia, C., & Piro, V. (2021). Ghettos, camps and dormitories: Migrant workers' living conditions in enclaves of industrial agriculture in Italy. In J. F. Rye & K. O'Reilly (Eds.), *International labour migration to Europe's rural regions* (pp. 52–69). Routledge.

Buier, N. (2024). "It is through struggle that we may write of structure". Agro-industrial crisis and Europe's workers without history. *Focaal—Journal of Global and Historical Anthropology, 100*, 40–53. https://doi.org/10.3167/fcl.2024.1000104

Caruso, F. S. (2018). Dal ghetto agli alberghi diffusi: l'inserimento abitativo dei braccianti stagionali nei contesti rurali dell'Europa meridionale. *Sociologia Urbana e Rurale, 116*, 78–92. https://doi.org/10.3280/sur2018-116006

Castillero, S. (2021). La esencialidad del cuarto mundo. 'Las marroquíes y los morenos' en la salvaguarda de la producción alimentaria en la frontera sur. *Revista de Estudios Sociales, 78*, 128–143. https://doi.org/10.7440/res78.2021.08

Ceccarelli, G., & Ciconte, F. (2018). *Human suffering in Italy's agricultural value chain*. Oxfam.

Chung, Y. B. (2017). Engendering the new enclosures: Development, involuntary resettlement and the struggles for social reproduction in coastal Tanzania. *Development and Change, 48*(1), 98–120. https://doi.org/10.1111/dech.12288

Corrado, A., & Caruso, F. C. (2022). Essential but exploitable: Migrant agri-food workers in Italy and Spain. *European Journal of Migration and Law, 24*(2), 193–216. https://doi.org/10.1163/15718166-12340126

Corrado, A., Caruso, F. S., Lo Cascio, M., Nori, M., Palumbo, L., & Triandafyllidou, A. (2018). *Is Italian agriculture a 'pull factor' for irregular migration – And, if so, why?* Open Society European Policy Institute.

De Castro, C. (2014). La Desdemocratización de Las Relaciones Laborales En Los Enclaves Globales de Producción Agrícola. In A. Pedreño Cánovas (Ed.), *De Cadenas, Migrantes y Jornaleros. Los Territorios Rurales En Las Cadenas Globales Agroalimentarias* (pp. 59–77). Talasa.

Dines, N. (2023). After entry: Humanitarian exploitation and migrant labour in the fields of Southern Italy. *Environment and Planning D: Society and Space, 40*(6), 1101–1118. https://doi.org/10.1177/02637758221117161

Echeverría Ramírez, M. C. (2003). *Hábitat versus vivienda: mirada crítica sobre el viviendismo*. https://repositorio.unal.edu.co/bitstream/handle/unal/6882/MCE02-HabVs-Viv.PDF

Echeverría Ramírez, M. C., Yory, C., Sanchez Ruíz, J., Gutiérrez Flórez, F., Zuleta Ruíz, F., & Muñoz Ciro, E. (2009). *¿Qué es el hábitat?: las preguntas por el hábitat*. Universidad Nacional de Colombia, Sede Medellín, Facultad de Arquitectura, Escuela del Hábitat - CEHAP.

Escrivà, A. (2022). Globalization and health: Gender issues in temporary agricultural work (Huelva). In N. Ribas-Mateos & S. Sassen (Eds.), *The Elgar companion to gender and global migration* (pp. 324–333). Edward Elgar Publishing. https://doi.org/10.4337/9781802201260.00036

Fanon, F. (1967). *Black skin, white masks*. Grove Press.

FitzSimmons, M. (1986). The new industrial agriculture: The regional integration of specialty crop production. *Economic Geography, 62*(4), 334–353. https://doi.org/10.2307/143829

Fravega, E. (2023). Looking for homes in migrants' informal settlements: A case study from Italy. In P. Boccagni & S. Bonfanti (Eds.), *Migration and domestic space* (pp. 153–170). Springer.

Fritsch, M., & Matthes, J. (2020). On the relevance of global value chains and the intra-European division of labour. *National Institute Economic Review, 252*, R4–18. https://doi.org/10.1017/nie.2020.16

Garcés, B., & Güell, B. (2021). Temporeros agrícolas en España: nuevas caras de un viejo problema. In J. Arango, B. Garcés, R. Mahía, & D. Moya (Eds.), *Inmigración en tiempos de Covid-19* (pp. 86–100). CIDOB.

Gialis, S., & Herod, A. (2014). Of steel and strawberries: Greek workers struggle against informal and flexible working arrangements during the crisis. *Geoforum, 57*, 138–149. https://doi.org/10.1016/j.geoforum.2014.08.014

Gudynas, E. (2010). Agropecuaria y nuevo extractivismo bajo los gobiernos progresistas de América del Sur. *Territorios, 5*, 37–54.

Gudynas, E. (2015). *Extractivismos: Ecología, economía y política de un modo de entender el desarrollo y la Naturaleza*. CEDIB.

Gudynas, E. (2016). Modos de producción y modos de apropiación: Una distinción a propósito de los extractivismos. *Actuel Marx / Intervenciones, 20*, 95–121.

Harnecker, M. (1971). *El capital: conceptos fundamentales. Manual de economía política*. Editorial Universitaria.

Harvey, D. (2010). *The enigma of capital and the crises of capitalism*. Oxford University Press.

INE. (1999). *Censo Agrario 1999.*

INE. (2009). *Censo Agrario 2009.*

INE. (2020). *Censo Agrario 2020.*

Istat. (2010). *Censimento Generale dell'Agricoltura 2010.*

Istat. (2020). *Censimento Generale dell'Agricoltura 2020.*

Kasimis, C., Papadopoulos, A. G., & Zografakis, S. (2015). The precarious status of migrant labour in Greece: Evidence from rural areas. In D. Della Porta, S. Hänninen, M. Siisiäinen, & T. Silvasti (Eds.), *The new social division: Making and unmaking precariousness* (pp. 101–119). Palgrave Macmillan UK. https://doi.org/10.1057/9781137509352_6

Kotsila, P., & Argüelles, L. (2023). The necropolitics of expendability: Migrant farm workers during COVID-19. *Journal of Peasant Studies, 51*(2), 441–465. https://doi.org/10.1080/03066150.2023.2243440

Kotsila, P., & Kallis, G. (2019). Biopolitics of public health and immigration in times of crisis: The malaria epidemic in Greece (2009–2014). *Geoforum, 106*, 223–233. https://doi.org/10.1016/j.geoforum.2019.08.019

Kukreja, R. (2021a). COVID-19, racial capitalism, and undocumented Bangladeshi agricultural workers in Manolada, Greece. *Dve Domovini / Two Homelands, 54*, 89–106. https://doi.org/10.3986/dd.2021.2.07

Kukreja, R. (2021b). Visible yet invisible: The disciplinary mechanism of self-surveillance among undocumented South Asian male migrants in rural Greece. *Journal of Ethnic and Migration Studies, 47*(15), 3486–3503. https://doi.org/10.1080/1369183X.2019.1642740

Kukreja, R. (2024). Undocumented migrant workers as activist citizens: Resisting citizenship exclusion through photovoice. *Citizenship Studies, 28*(4–5), 403–423. https://doi.org/10.1080/13621025.2024.2425646

Lefebvre, H. (1970). *La révolution urbaine*. Gallimard.

Lefebvre, H. (1974). *La production de l'espace*. Anthropos.

Macciani, C. (2024). *Oppression and resistance of racialized migrant farmworkers in the context of agricultural racial capitalism: The case of West African farmworkers in the province of Foggia (Italy)* [Doctoral dissertation, Università della Calabria].

Martínez-Valderrama, J., Gartzia, R., Olcina, J., Guirado, E., Ibáñez, J., & Maestre, F. T. (2023). Uberizing agriculture in drylands: A few enriched, everyone endangered. *Water Resources Management, 38*, 193–214. https://doi.org/10.1007/s11269-023-03663-1

Marx, K. (1867). *Capital: A critique of political economy. Volume I: The process of production of capital*. Progress Publishers.

McKay, B. M., & Colque, G. (2022). Dinámicas extractivas en el sector agrario de Bolivia. In S. M. Borras Jr. & J. Franco (Eds.), *Extractivismo agrario en América Latina: Caminando por las sendas de lo común para confrontar el acaparamiento de tierras, la agroindustria y la financiarización* (pp. 43–64). CLACSO.

McMichael, P. (2021). Political economy of the global food and agriculture system. In A. Kassam & L. Kassam (Eds.), *Rethinking food and agriculture. New ways forward*. Woodhead Publishing.

Melossi, E. (2021). 'Ghetto tomatoes' and 'taxi drivers': The exploitation and control of Sub-Saharan African migrant tomato pickers in Puglia, Southern Italy. *Journal of Rural Studies, 88*, 491–499. https://doi.org/10.1016/j.jrurstud.2021.04.009

Mies, M. (1986). *Patriarchy and accumulation on a world scale. Women in the international division of labour*. Zed Books.

Molinero-Gerbeau, Y. (2020). La creciente dependencia de mano de obra migrante para tareas agrícolas en el centro global. Una perspectiva comparada. *Estudios Geográficos, 81*(288), 1–27. https://doi.org/10.3989/estgeogr.202046.026

Molinero-Gerbeau, Y. (2021). The problem is not COVID-19, it's the model! Industrial agriculture and migrant farm labour in the EU. *EuroChoices, 20*(3), 69–74. https://doi.org/10.1111/1746-692X.12308

Molinero Gerbeau, Y., & Avallone, G. (2016). Producing cheap food and labour: Migrations and agriculture in the capitalistic world-ecology. Social Change Review, 14(2), 121–148. https://doi.org/10.1515/scr-2016-0025

Molinero-Gerbeau, Y., López-Sala, A., & Şerban, M. (2021). On the social sustainability of industrial agriculture dependent on migrant workers. Romanian Workers in Spain's Seasonal Agriculture. In Sustainability (Vol. 13, Issue 3). https://doi.org/10.3390/su13031062

Moore, J. W. (2010). The end of the road? Agricultural revolutions in the capitalist world-ecology, 1450–2010. *Journal of Agrarian Change, 10*(3), 389–413. https://doi.org/10.1111/j.1471-0366.2010.00276.x

Moore, J. W. (2015). *Capitalism in the web of life: Ecology and the accumulation of capital.* Verso.

Moore, J. W. (2016). The rise of cheap nature. In J. W. Moore (Ed.), *Anthropocene or Capitalocene? nature, history, and the crisis of capitalism* (pp. 78–115). PM Press.

Moore, J. W. (2017). Metabolic Rift or Metabolic Shift? Dialectics, Nature, and the World-Historical Method. *Theory and Society, 46*(4), 285–318. https://doi.org/10.1007/s11186-017-9290-6

Moore, J. W. (2022). Imperialism, with and without cheap nature: Climate crises, world wars and the ecology of liberation. *Working Papers in World-Ecology,* 3/2022. World-Ecology Research Group, Binghamton University.

Moraes, N., Gadea, M. E., Pedreño, A., & De Castro, C. (2012). Enclaves globales agrícolas y migraciones de trabajo: Convergencias globales y regulaciones transnacionales. *Política y Sociedad, 49*(1), 13–34.

Mozo, C., Moreno-Nieto, J., & Reigada, A. (2022). Health as an emerging value in the construction of quality of berries in Huelva (Spain). *International Sociology, 37*(6), 698–715. https://doi.org/10.1177/02685809221115960

Murrey, A., & Mollet, S. (2023). Extraction is not a metaphor: Decolonial and Black Geographies against the gendered and embodied violence of extractive logics. *Transactions of the Institute of British Geographers, 48*(2), 446–461. https://doi.org/10.1111/tran.12610

Newsham, N., & Rowe, F. (2023). Understanding trajectories of population decline across rural and urban Europe: A sequence analysis. *Population, Space and Place, 29*(3), e2630. https://doi.org/10.1002/psp.2630

Nori, N., & Farinella, D. (2020). Mobility and migrations in the rural areas of Mediterranean EU countries. In M. Nori & D. Farinella (Eds.), *Migration, agriculture and rural development* (pp. 39–67). Springer. https://doi.org/10.1007/978-3-030-42863-1_3

Papadopoulos, A. G., & Fratsea, L. M. (2021). Migrant and Refugee impact on well-being in rural areas: Reframing rural development challenges in Greece. *Frontiers in Sociology, 6*, 592750. https://doi.org/10.3389/fsoc.2021.592750

Papadopoulos, A. G., & Fratsea, L. M. (2024). 'No man's land': Reflecting on and theorizing migrant labour in the Mediterranean agriculture. In R. Zapata-Barrero & I. Awad (Eds.), *Migrations in the Mediterranean* (pp. 383–401). Springer. https://doi.org/10.1007/978-3-031-42264-5_22

Papadopoulos, A. G., Fratsea, L. M., Spyrellis, S., & Baltas, P. (2021). Exploring the contribution of migrant labour in Greek agriculture. *Italian Review of Agricultural Economics (REA), 76*(1), 33–48. https://doi.org/10.36253/rea-12574

Pedreño Cánovas, A., Sánchez-García, M. A., Gadea-Montesinos, M. E., & De Castro, D. (2022). Purity and danger in Spanish agriculture: Farm workers during the pandemic. *Estudios Geográficos, 83*(293), e108. https://doi.org/10.3989/estgeogr.2022114.114

Raeymaekers, T. (2024). Race, space, and social reproduction: The spatial organization of racialized labour in Italian food agriculture. *Environment and Planning D: Society & Space, 42*(3), 1–22. https://doi.org/10.1177/02637758241287339

Reigada, A. (2022). *Historia, trabajo y territorio. El conflicto capital-vida en los campos de fresas de Huelva*. Publicacions i Edicions de la Universitat de Barcelona.

Rodrigues de Azevedo, F. T. (2023a). Geografias do não ser: efeitos psíquicos do racismo na produção do espaço. *Revista da ABPN, 16*, 928–962. https://doi.org/10.31418/2177-2770.2023.v16.c.1

Rodrigues de Azevedo, F. T. (2023b, May 22–26). *Pele negra, medo branco: sobre interdições e o não-ser metropolitano*. [Conference session]. XX ENANPUR 2023, Belém, Brasil. https://anpur.org.br/wp-content/uploads/2023/05/st12-18.pdf

Rodríguez-Calles, L. (2022). Ecological impacts of agribusiness transformation in a Spanish Mediterranean enclave: Impacts on the Mar Menor coastal lagoon. *EuroChoices, 21*(2), 43–49. https://doi.org/10.1111/1746-692X.12362

Rodríguez-Calles, L. (2023). *Las actitudes del racismo y las relaciones de convivencia intergrupales. El caso de Torre-Pacheco (Murcia)* [Doctoral dissertation, Universidad Pontificia Comillas]

Rodríguez-Calles, L, & Iglesias, J. (2025). Factores que explican la coexistencia tranquila entre inmigrantes y nativos en barrios populares españoles. Los casos de Arangoiti (Bilbao) y Torre-Pacheco (Murcia). *Revista Migraciones*, 1–22. https://doi.org/10.14422/mig.2025.001

Rodríguez-Guerrero, L. A., Pérez-Urdiales, I., Escrig-Piñol, A., Jiménez-Lasserrotte, M. M., Pastor-Bravo, M. M., Mateos, J. T., & y Briones-Vozmediano, E. (2024). Water Insecurity among seasonal agriculture workers: Perspectives from Spanish Professionals. *International Journal for Equity in Health, 23*(1), 31. https://doi.org/10.1186/s12939-024-02112-8

Rubio González, M., Jiménez-Lasserrotte, M. M., Ugarte-Gurrutxaga, M. I., El Faqyr, K. E. M., Granero-Molina, J., Fernández-Sola, C., & Plaza del Pino, F. M. (2023). Experiences of the migrant farmworkers during the syndemic Due to COVID-19 in Spain. *Social Sciences, 12*(5), 273. https://doi.org/10.3390/socsci12050273

Ruíz Ramírez, C., Castillo-Rojas-Marcos, J., & Molinero-Gerbeau, J. (2024). *Essential but invisible and exploited: A literature review of migrant workers' experiences in European agriculture*. Oxfam-Intermón.

Sajir, Z., Molinero-Gerbeau, Y., & Avallone, G. (2022). "Todo cambia, todo sigue igual". La gobernanza de la mano de obra migrante en la agricultura española e italiana en el primer año de la pandemia de COVID-19. *Estudios Geográficos, 83*(293), e120. https://doi.org/10.3989/estgeogr.2022120.120

Salvia, L. (2020). The restructuring of Italian agriculture and its impact upon capital–labour relations: Labour contracting and exploitation in the fresh fruit and vegetable supply chain of the Lazio Region, Central Italy. *Journal of Agrarian Change, 20*(1), 98–112. https://doi.org/10.1111/joac.12340

Sayad, A. (1984). État, nation et immigration: L'ordre national à l'épreuve de l'immigration. *Peuples Méditerranéens, 27–28*, 187–205.

Smith, N. (1984). *Uneven development: Nature, capital and the production of space*. Blackwell.

Svampa, M. (2013). Consenso de los commodities y lenguajes de valoración en América Latina. *Nueva Sociedad, 244*, 30–46.

Wallerstein, I. (2010). Free flows and real obstacles. Who wants Laissez-Faire? In T. A. Jones & E. Mielants (Eds.), *Mass migration in the world-system: Past, present and future* (pp. 13–21). Paradigm Publishers.

Zapata Rivera, R. V. (2022). Luchas Sindicales En La Agricultura Californiana, En La Década de Los Treinta. *Secuencia*. https://doi.org/10.18234/secuencia.v0i112.1813

AGRICULTURAL PROJECTS FOR A SUSTAINABLE FUTURE: VISIONS AND CONTRADICTIONS

Nature, Capital, and Control in the Organization and Digitalization of Pig Production

Irene Guerrero Avellaneda, *Andrés Pedreño Cánovas*, *and Antonio José Ramírez Melgarejo*

Introduction

The Spanish Government's Ministry of Agriculture, Fisheries and Food annually publishes the data from a livestock slaughtering sector survey. Looking back , from 2014 to 2015 we witness a noteworthy leap (of over one million pigs, or a 4.15% increase) in the pig population compared to the relatively stable figure for the 2007–2013 period (around 25–26 million pigs). In 2020, the year of the pandemic, the pig population experienced a spectacular increase, reaching a record number of 56.4 million

I. Guerrero Avellaneda (✉) • A. Pedreño Cánovas
Department of Sociology, University of Murcia, Murcia, Spain
e-mail: irene.guerreroa@um.es; andrespe@um.es

A. J. Ramírez Melgarejo
Department of Applied Sociology, Complutense University of Madrid, Madrid, Spain
e-mail: antramir@ucm.es

S. Narotzky et al. (eds.), *Agricultural Extractivism in the Mediterranean Region*,
https://doi.org/10.1007/978-3-032-05599-6_8

animals slaughtered (a 6.5% increase compared to the previous year) and some 5 million tons of meat produced (a 8.2% increase). Spain then positioned itself internationally in the corporate food regime (McMichael, 2016) as a global area of specialization, becoming fourth in world pig meat production behind China, the United States, and very close to Germany which totals 22% of EU production, compared to 21.8% Spanish production. Yet, it has recently surpassed Germany in number of animals sacrificed (MAPA, 2023) and, despite a drop in production, Spain remains the number one producer in the EU and the third globally (MAPA, 2023), only behind China and the United States.

This increase in intensive pig production entailed organizing an extractive logic of resources (water, soil, animals) and labor, while also involving the creation of health risks (animal diseases, contagion, and zoonosis) and environmental risks (contamination due to water and soil slurry, unpleasant odors, overuse of aquifers), that alerted the population and triggered the mobilization of many rural towns against the location of industrial pig farming in their areas.

Institutional and corporate players that promote pig industrialization have presented digitalization as a technological solution to the health and environmental problems caused by intensification and nature extractivism. On the other hand, public food safety, animal welfare, and environmental regulations have promoted measures to fight environmental and sanitary risks but have failed to question the organizational model of industrial pig farming.

However, social agribusiness studies have developed a notable tradition of critical research on the sustainability of the agribusiness system. David Goodman recently proposed that the convergence of digital and molecular technologies[1] represents an evolution, rather than a revolution, of the agribusiness paradigm. In fact, as Goodman shows, digital and molecular technologies have become integrated into the classic agribusiness logic, without it entailing a transition to eco-sustainable production (Goodman, 2023). Another line of research exposes the "illusion of control" that is deposited in technological changes in both intensive agriculture (Stuart, 2008) and intensive animal farming (Worosz et al., 2008).

[1] It refers to the evolution that aims to integrate digital technologies molecular processes (or biotechnological), meaning the biology and digital worlds are no longer separate fields and are becoming parts of an interconnected technological ecosystem.

In this chapter we present the workings of the technological shift towards digitalization in pig production. We posit that digitalization reinforces the logic of extraction of nature and labor in the pig farming industry. This is because it entails continuity and a push for production specialization and intensification dynamics, as well as business centralization and concentration. Likewise, digitalization promises to control environmental, health and hygiene risks, which is a key source of legitimacy for its expansion. However, this narrative of control may end up being an illusion, as long as the extractivist logic remains in the pig farming industry.

The chapter is divided into the following sections. First, we present the research conducted in the Murcia Region on the organizational model of the pig production commodity chain and its territorial and extractive dimensions, wherein digitalization is strategic as a means of coordination (Section "A Chain Across the Territory"). Next, follows an analysis of the managers' and producers' view of digitalization as a narrative of control and expected salvation (Section "Digitalization and Risk Control"). The last section analyzes the transformations of production relationships resulting from this new production system. We conclude by addressing the theoretical contributions of the chapter.

A Chain Across the Territory

This chapter is based on research conducted in one of the largest production regions in Spain, the Region of Murcia. This autonomous territory is home to around 7% of national pig production, despite representing only 2.24% of the country's surface area. For the purposes of this research, a methodological strategy was designed that combined different techniques. On the one hand, we employed a qualitative approach based on 25 in-depth interviews and two focus groups, structured according to differentiated profiles, reflecting participants' relationship with the type of production channel (see Table 1). On the other hand, we adopted a sociological approach, collecting data through direct intermittent observation of the territory over a two-year period. Numerous informal conversations were held with workers and farmers. In parallel, our formal and casual meetings with local residents opposed to the operations of large-scale farms facilitated our guided visits to the affected territorie and provided valuable information. Finally, the authors have been living in and conducting research on agro-industry in this region continuously for over a decade. And one of us maintains a strong network of interlocutors in the

Table 1 Summary of interviews and focus groups carried out in the pig sector, Murcia Region

Entity/profile	*Interview and focus group codes by job title/function*
Large meat processing companies	Assistant to the management (E1); Human Resources Manager (E6)
Integrated pig farms	Group Leader (E4), Farmer (E8, E11, E15, E20, E21)
Short production channel (extensive and organic), processing and marketing	Farmer (E16)
Large Meat Processing Company Workers	Worker in the production chain (E5, E7, E18, E22), GD1 (Focus Group with Representatives of the Works Council from a Large Meat Processing Plant)
Family run slaughterhouse	Manager (E14)
Service companies	Valorization of pig slurry (E2), digitization (E10, E12), environmental consultancy (E13), veterinary (E13, E24)
Consumers	Representative of the organic consumption association (E19)
Activists against the effects of pig farms	Platform for the defense of sources (E9), rural development associative entity (E17, E23, E25), focus group with affected neighbors (GD2)

town where the large meat processing plant is located. This has allowed us to sustain close, daily contact with many of the interviewees and to closely follow - and become immersed in- the social dynamics under investigation.

The research focused on two large meat production companies whose central activity is located in the Guadalentín Valley (Murcia Region). These large meat companies, running their own food distribution brands, are the central core of the integrated pork production chain.

On the one hand, we studied a large company specializing in the livestock production phase (hereafter G1) located in Lorca, in the Murcia Region. This company emerged as part of the "feed revolution" that the town experienced in the 1960s and 1970s, a time when feedlots expanded as spaces for intensive exploitation of the pig-fattening or finishing cycle, to the detriment of traditional piglet production. This shift marked a fundamental step toward the integration of the sector into the meat industry (Segura et al., 1991). Today, G1 consists of over 600 farms employing around 400–450 wage laborers, with a total annual output of two million pigs. Approximately half of this production goes to a slaughterhouse owned by the same company, while the rest is sold to slaughterhouses across the country (one of the closest clients is the G2 slaughterhouse).

On the other hand, we studied company G2, a conglomerate that encompasses the livestock farming phase, the industrial meat processing phase, and the commercial phase through its proprietary brand. This company defines its own organization as "a unique circular management model" called Integrated Process Control (IPC), which allows it to control all phases of the production process.

The pig production model is similar in both G1 and G2. These are individual farms employing wage laborers, though they also resort to the vertical farm integration model, a widespread hiring formula in the pig sector (covering 75% of farms, according to COAG, 2019). This type of arrangement is based on a "contractual livestock relationship in which one party, called the integrator [both G1 and G2], is obligated to provide animals and/or products for animal feed, health and veterinary assistance products, while the other, called the integrated livestock farmer (self-employed farmer), provides livestock housing services, facilities, labor force and animal care" (Law 8/2003 on Animal Health).

In recent years, as part of the vertical integration process, traditional methods rooted in experience, trade knowledge, and analog automation tools have been replaced with technological devices for production control.

Both G1 and G2 paradigmatically represent the change that large meat companies and food distributors underwent at the turn of the century, when they began creating their own brands, positioning themselves at the center of the integrated pork chain. This process entailed profound organizational changes, as brand ownership enabled greater control and vertical integration over producers, who had to adapt their processes to comply with quality standards, delivery terms, and demands for greater product uniformity (Soldevila, 2006, p. 715).

Therefore, both companies experienced a restructuring process toward a "network organization model" (Pedreño et al., 2021), a structure increasingly adopted by agrifood companies under the current global corporate regime (McMichael, 2016) to create "flexible production ecosystems" (Cattaneo & Bocchicchion, 2019). This model enables efficient adaptation to market fluctuations and operates through strong cooperation and exchange between producers, suppliers, and buyers. Beyond flexibility, this network organization allows the central company to define and control the production timelines for all the actors involved in production and distribution, thanks to the implementation of new digital devices managing all the production and distribution stages, from live animal breeding to meat products.

Therefore, with this organizational structure, the relationships between large companies and other actors (wage laborers, integrated livestock farmers, external workers, outsourced services, etc.) follow a logic of domination and control, as the central company dictates and implements production rhythms. Moreover, in the vertical integration process, the growing investment in technological management tools have progressivly replaced traditional methods rooted in experience, trade knowledge and farmers' tacit know-how.

For both G1 and G2, we observed an intensification in the subsumption of producers (both individual and integrated farmers), through more thorough and detailed control mechanisms and a stronger emphasis on vertical production submission, aimed at aligning production with quality standards, tight delivery deadlines, and the pursuit of uniform products (Soldevila, 2006, p. 715).

This network organization represents a logic of productivism that requires extensive use of space for the dispersed distribution of farms and livestock operations. Based on the constant movement of goods, via hundreds of trucks transporting animals, consumables, and finished products (such as electronics) through Information and Communication Technologies (ITC), the pig livestock integrating companies operate as massive logistics operators. They control the production process both vertically and horizontally, from the genetic selection of the animals (though purchasing piglets from third-party European countries is also common) to the distribution of finished products on international supermarket shelves.

In this network model, farms are highly specialized in livestock farrowing and fattening (see Table 2).

This breakdown of production tasks and the resulting specialization of farms forms a network within which lies the center of coordination and control for the meat plant. This system relies on an organizational model that manages both space and nature accross a vast rural geography hosting thousands of production farms, all coordinated from the organizational core located in the Murcia Region. Both G1 and G2 form a geography

Table 2 Life cycle of a pig

Phase	*Number of kilos*	*Duration of each phase*
PHASE 1: Sow farm	1 kg–6 kg	Sows' gestation cycle 4 and a half months Piglet farrowing: 15 days
PHASE 2: Weaning	6 kg–20 kg	2 months
PHASE 3: Fattening and finishing	20 kg–115/150 kg	4 months/4 and a half months

Own elaboration based on interviews with GE managers and farmers

mainly found across the southern half of Spain,[2] although other regions, such as Castille-León or the Valencian Community, may also be included.

Digitalization is a key factor in the transformation of these production enclaves into agrifood logistics management centers. Today, in addition to coordinating their own production, they manage the manufacture of by-products, adding value to waste, such as manure made from slurry or animal feed products. In line with the network logic, slaughterhouses and integrating companies increasingly exchange both fattened animals and finished products according to production and demand fluctuations.[3]

There is a logic of production of nature implicit in this way of organizing production and extraction of natural resources, essential to the production process. If farms are segmented into specialized tasks and widely distributed, so too are the risks of disease transmission, environmental pollution (unpleasant odors, excreted antibiotics, and slurry), and natural resource extraction (soil, water, and atmosphere). This arrangment aims to distribute or disperse the hygiene and socio-environmental risks of the production process in order to minimize their impact, comply with

[2] The location of the five G2-owned factories for the production of its own compound feed is an example of this southern geography: two in Lorca (Murcia Region), one in Jaén (Andalusia), one in Talavera (Castilla-La Mancha), and another in Toledo (Castilla-La Mancha).

[3] For example, these fluctuations may occur due to a seasonal drop in production, as animals' metabolisms may slow down during temperature spikes, which regularly occur in southern Spain, resulting in the fattening phase taking some weeks longer than usual. Likewise, this may be caused by increased demand during holidays when more ham is consumed, such as during the Christmas period. One common response to these fluctuations is the buying and selling of animals and carcasses between G1 and G2 and other brands, supermarket chains, and slaughterhouses.

environmental waste management regulations and reduce possible social responses and/or large neighborhood protests.

The geographical extension and dispersion of the networking group require time-space compression (Harvey, 1990), achieved through the rational organization of social time, replacing space with intensive and efficient time considerations. This involves the specialization and centralization of production, the deployment of a logistics transport network, the digitalization of production processes, the streamlining of work schedules, and the reduction of animal lifespans.

Specialization, Concentration, and Centralization of Livestock Farms

With increasing production scales and product standardization as core requirements for competitiveness in the global and corporate food regime, the pork sector has moved toward greater pig-farm concentration. The so-called "factory farms" exemplify the time-space compression process of recent decades: the number of pigs and farms has notably increased at the cost of eliminating small- and medium-sized farmers, who are now integrated as producers into corporate groups that concentrate control over a growing number of farms.

According to the 2023 agricultural census, the sector has experienced substantial restructuring over the past 19 years. The total number of farms has significantly decreased, particularly among the smallest livestock operations (also called "reduced farms," with a maximum capacity of 42 animals), which have declined by nealry 27% over the last decade. Meanwhile, larger farming operations have increased by 42% in the last 10 years, particularly those belonging to group 3 (480–720 Unidad de Ganado Mayor (UGM, Large Livestock Unit) or farms with up to 6000 finishing pigs in phase 2 or up to 36,000 piglets in phase 1) (MAPA, 2023).

A Vast, Specialized Road Transport Network

A central element in the coordination of the production network is a specialized road transport, comprising trucks called "cages" for transporting livestock between farms covering different phases and the slaughterhouse. Truck drivers are typically outsourced as self-employed workers. The organization of livestock transport is the main centripetal force that counteracts spatial dispersion, substituting space with intensive labor. High demand and strict schedules, hygiene and welfare regulations and winding, rural roads, make transportation management a tense issue. The potential for accidents, some of which may be fatal, is very high, considering this is a moving cargo (live animals) transported by often exausted drivers traveling many kilometers on rough country roads.

Digitalization of Processes

Digitalization enables the central organizational unit to precisely coordinate and control farms across the territory. At the same time, it is also a form of substitution of space for time, as the manager of the large meat company (G1) explained in an interview: "*Our 600 farms are located across all of Spain, so they are very far apart. This is not a very local production center… No, no, in the southern half of the peninsula we have farms in almost all the provinces. So, farmers have mobile applications at their facilities that they use to enter a whole lot of data. They have an application that orders the feed they need and log data there of how much feed has been used*" (E1).

Digitalization is applied to all production processes, especially via the Internet of Things (IoT) for data collection and analysis at the central organizational unit (in phase 0 and own farms) and via the sensorization of spaces, in order to control factors such as humidity, temperature, or the presence of gases (especially in phase 0, and decreasing as the pigs gain weight in phases 1 and 2). Sensorization aims to increase monitoring and reduce polluting emissions, which, beyond the environmental and animal welfare discourse, seeks to increase the livestock operation and the appropriation of human and non-human nature. (Moore, 2020)

On the other hand, digitalization exacerbates inequality within the sector, inhibiting the sustainability of less capitalized producers, whether integrated or not, favoring instead production concentration in favor of efficiency.

Streamlining Labor Time

Farm intensification and specialization has digitized tasks so that a small number of employees can manage a large number of farms with considerable production volume. Such is the case of the G2 farm we visited in Cancarix (Hellín, Albacete) --one of the largest factory farms in Spain with 12 fattening units of 6000–7000 places each (divided into four sheds with 1200–1500 pigs each) and four piglet units. In G2, a single "unit supervisor" oversees the tasks of one finishing unit (i.e., 6000 pigs), involving monitoring feeding, vaccination and general animal care. In total, there are 70 workers on the farms and 10–15 tractor drivers who remove slurry for an estimated production of 120,000 fattener pigs or 120,00 livestock spaces.[4]

This streamlined workforce seems to fulfill the corporate dream of a worker-free farm. But this ideal is disproved, not only because farms still require human labor such as for supervision, veterinary care, and manual tasks (cleaning, animal handling, etc.), but above all for truck-based transport of live livestock and other inputs (feed, slurry, foodstuffs, water tanks, etc.) following the capitalist logic of production delocalization.

Given this persistence of human labor need, the valorization of capital finds other ways to reduce costs, by outsourcing tasks to falsely self-employed workers, hiring cheapened immigrant labor, and extending working hours, among others (Pedreño et al., 2021).

Vertical integration contracts for independent farmers predominate over self-employed farm models, efficiently shifting risk to the integrated farmer. The integrated farmer, not only acquires the land,[5] finances the construction and maintenance of the facilities, bears the loss of profits due to the death of animals, and provides the workforce to the integrating company without direct compensation (because they are a self-employed worker), but is also the one absorbing the socioecological costs related

[4] All the data were extracted from the E8 interview with the G2 integrated farm manager.

[5] The land where integrated farms are located is usually owned by farmers and is either inherited or acquired by fulfilling the conditions required by the regulations governing the setting up of farms (Royal Decree 206/2020) (distance from urban centres, water resources, no protected species nearby, close to agricultural operations for dumping slurry, etc.). It is fairly common for integrating companies to act as guarantors for farmers when the latter apply for a loan to build farms, especially when integrating companies have been wanting to expand their production capacity but have not been willing to assume the costs involved in starting up the farms. Integrating companies either lend the money or provide collateral so farmers can apply for a loan from the bank. In case of non-payment, the integrating company keeps the farm.

This is another mechanism that emphasizes the imbalance between the two contractual parties since the positive economic performance of the farm—and therefore the payment of fees—depends on the integrating company's willingness to continue providing the integrated farm with animals to care for. The contracts between the two parties are renewed with each shipment of pigs that enters the farm, every three or four months or so, so continued activity is not guaranteed for the following years until the investment is recouped.

to land and water hoarding, bad odors, environmental contamination, and the public discontent they bring along.

During our fieldwork, we visited several integrated farms that had been set up in recent years in the municipality of Socovos, in the Sierra del Segura region of Albacete. Those dependent self-employed farmers subsumed or literally "integrated" into the logic of large industry, predominantly bear such costs without benefitting from the surplus value that their work generates. By contrast, their partnership with larger companies ensures the later's feed supply and competitive sales in a context of economies of scale.

Shortening the Animals' Life Cycle

Animal lifespans are significantly compressed following the intensive logic of producing more in as little time as possible. Piglets reach optimal weight (115–120 kgs -a weight gain of approximately 800 grams per day) in just six months. When they "are ready" (E8), a fleet of permanently available trucks transports them to the slaughterhouse for producing cured meats or sausages in just 24/48 hours.

This space-time compression is achieved by 1) Constantly seeking to increase the hyperproliferation of gestating sows via genetic selection and improved farm conditions[6] and 2) feed efficiency, in other words, reducing "*the amount of plant-based protein needed to create animal protein*" (E1). The efficiency level is measured using a feed conversion ratio calculated by dividing the kgs of feed needed per kg finished pig:

> This used to be around 5, meaning you needed 5 kilos of feed to produce one kilo of pork, now it's around 2.5 or less. 2.5 kilos of feed to produce 1 kilo of pork... We produce over 600,000 tons of feed and produce 200 and some thousand tons of meat per year. So those two million pigs weigh 200,000 tons and change. To produce those 2 million pigs that weight around 200,000 tons, we needed around 600,000 tons of feed. (E1, GE manager interview, November 22, 2023)

[6] Sow performance is an important production cost, leading countries to compete in increasing their productivity as much as possible: "Denmark and the Netherlands lead the sow productivity list with about 28.0 weaned pigs sold per sow per year. They are followed by Germany, Belgium, France, and Ireland, with 25–26 pigs sold per female per year. Meanwhile, most other European and American countries produce 22 to 24 weaned pigs per female per year" (Hoste, 2020).

Digitalization and Risk Control

On October 24, 2023, we visited the 56th annual Livestock, Industrial and Agrifood Fair, SEPOR, held in the city of Lorca, a core town for regional pig porduction. SEPOR is an event that brings together pig producers from the Region of Murcia, featuring technical conferences, exhibitions of technological and organizational innovations, and supplier centers, among others.The first thing that caught our attention was the poster announcing that year's fair.

The poster design featured a cyborg pig illustration, whose insides, made up of turbines, pulleys, solar panels and cables, looked more like a factory than a living being. The pig's gaze was lacking in expression, as if it had been hypnotized and could no longer control its own actions or perceive the world. Both the poster and the fair activities were tinged with a feeling of techno-optimism (Barry, 2012). It was as if that cyborg animal was linked to an immaculate, smooth, shaved, and abstract nature associated with an ideal of sustainability thanks to solar panels, wind energy, and woodlands. There was no trace of human labor or extraction of nature, as if the event's slogan "feeding the future" had fulfilled the dream of the technological elite to ultimately eliminate people and nature from the production process. The illustration shows "the illusion of control" of nature and labor (Stuart, 2008) that the pig sector promotes today.

In corporate discourse, the digitalization of farms is presented as the meticulous measurement of every single element involved in the various processes that make up the production chain, in which human labor is made invisible. In this way it contrasts with their so-called "clinical eye." As one of the human senses, the human eye is considered to be imprecise and prone to error when detecting, for example, if a pig is underfed or if an infectious virus has entered the drove. Digital technologies, on the other hand, make it possible to predict processes and reduce uncertainties and, therefore, risks. Health risks can be detected more quickly, resulting in greater efficiency in minimizing animal losses or the possibility of infection:

> Well, the systems are becoming more and more intelligent, I mean "if I have a feeding pattern of so many kilos per day, the animals… I can't have 1,000 animals and use up one thousand tons today and tomorrow, with those same thousand animals, use up 700," something isn't right. They are also used in the process of extracting information so we can analyze the data to see how

production is going. (E12, interview with an engineer responsible for designing digital devices, digital services company, March 01, 2024)

It's true that we react much faster. As far as diseases that appear on the farm, before we might see a spike in abortions and we could say "abortions are up" on the farm… so you know where we're heading, right? But we had to wait 15 days to actually see it in the program, when now we can instantly see what is happening that day. But I don't know if the incidence has gone down because of that, because on the farms, handling is also much more important, the handling of the animals, with biosecurity…well, that depends on what it includes. (E3, interview with veterinarian, December 01, 2023)

In recent decades, public policy has placed significant emphasis on food safety regulations. The goal has been to create a specific, pristine, and sealed-off environment to prevent zoonotic diseases and viral transmission among animals. Digitalization has been central to realizing this idea of the farm: "We isolate the pigs from the outside world and determine everything that happens inside: temperature, humidity, when it's daytime or night-time, when they eat or don't eat, whether they breathe or don't breathe… we control absolutely everything" (E12, March 01, 2024). Other measures have focused on modifying animal handling or improving farm conditions - particularly in lactating sow farms- to enhance pigs' quality of life.

These measures have led to a further production concentration/centra lization in factory farms and accelerated digitalization. For instance, animal welfare legislation has restricted the indiscriminate use of antibiotics and other drugs to treat pig diseases. This has, in turn, led to a greater digitalization of farms in order to,prevent diseases by exerting more control over the quality of feed or by ensuring the quantity of food that each pig eats. Additionally, these digital systems allow for a quick detection of disease onset. Affected animals are promptly removed from the production cycle and isolated in a dedicated treatment pens.

Nevertheless, there is still resistance to introducing such digital innovations on farms. Technicians and engineers report that younger farmers are more inclined to adopt new technologies compared to older, more traditional farmers. However, the material conditions of life are decisive and rank livestock farmers by their ability to invest in these technology systems. Digitalization is taking hold in larger, better capitalized farms, while smaller farmers who cannot afford such costly investments continue to refer to on-the-job experience and their "clinical eye."

Large corporate-owned farms can bear the cost of digitalization, adapting their facilities to minimize health risks. Likewise, large integrated farms with sufficient capitalization can also take on such investments. In contrast, small integrated farmers end up on the losing end, as they lack sufficient investment capacity. Their integration contracts with large companies do not support their technological transition.

Farm modernization and digitalization are also associated with the ability to control environmental risks. One key concern is the treatment of slurry to minimize its polluting emissions in the atmosphere and aquifers. Measures include waterproofing slurry disposal pits, frequent tank emptying, covering slurry with plastic sheeting instead of leaving it exposed, banning fanning, and improving the handling of manure. Digitalization is also deployed as a means to monitor and control these processes:

> Not yet in Murcia, but they already have it in Catalonia. Tractors have GPS, the tractor that transports the slurry has GPS and measures where it is spread, and it has a flow meter; GPS and a flow meter, it's mandatory. And in Murcia it will be mandatory soon as well… And not yet in Castilla-La Mancha, it's not mandatory either. They started in Catalonia and now they're heading towards Aragon, the areas that were most contaminated… You know. So, ultimately, they have to jump on the digitalization bandwagon. (E13, environmental advisor, worked for years on a G2 factory farm, March 04, 2024)

The Mar Menor[7] crisis, caused by nitrate pollution, exposed the workings of 446 intensive pig farms in the watershed, with an approximate census of eight hundred thousand heads, mostly concentrated in a single municipality. These farms produced environmental damage due to inadequate treatment of nitrate and antibiotic waste, both in terms of storage and dispersal on agricultural lands. In addition to the slurry pit treatment measures, in the Mar Menor watershed, the law requires that livestock

[7] The Mar Menor is a coastal saltwater lagoon in the region of Murcia, in south-east Spain, that has suffered severe environmental damage after nitrates and phosphates from intensive agriculture and livestock farming washed into it.

farmers must fill-in the REMODEGA registry (Electronic Registry of Livestock Manure Movements: manure and slurry)[8]:

> It's an electronic system that controls the amount of slurry you produce. You send it to a… receiver, and the receiver also makes a report, and the producer's report has to match that of the receiver… [The factory farm] already has programs, I used to work there, it has computer programs that control how much slurry is dumped onto each plot. Even if the tractor driver goes over, it lights up red, it's a visual thing, a SIGPAC map… it lights up red if you dump too much, green if you're okay, yellow if they're about to reach the limit. All of these are new technologies for tractor drivers, they take time to learn, but it's just a matter of getting used to it, because until now, we worked a bit blindly and just guesstimated. Until now, we didn't do this with slurry, but now we'll have to. If I have a tractor that spreads five cubic meters per minute, and I've spread twenty cubic meters twenty minutes… They don't only have a flow meter, some have…, you can't analyze the nitrogen that quickly, but they have something similar, a conductivity meter, they measure the electrical conductivity of the slurry. So, in terms of conductivity, which is easy to measure, you know if the slurry is more or less rich, and if you have to go faster or slower with the tractor. I can go a bit slower with slurry from sows because it has less nitrogen, but if I go with slurry from finishers, I have to go faster. (E13, March 04, 2024)

ECOGAN is a computer system developed by the Spanish government that records Best Available Techniques (BATs) applied on the farm and also estimates pollutant and greenhouse gas emissions and resource use on a specific farm throughout the production process.[9] This legislation differentiates between "farms with modern technologies," which are required to reduce their emissions by 80%, and "traditional farms," which are required to reduce their emissions by 40%, according to article 16.2 of Royal Decree 306/220.

Regarding groundwater extraction for farming, digitalization is also used in business discourse as a guarantee of sustainable water use, measuring the precise animal water need and use:

[8] See https://sede.carm.es/web/pagina?IDCONTENIDO=3653&IDTIPO=240&RASTRO=c$m40293

[9] https://www.mapa.gob.es/es/ganaderia/temas/ganaderia-y-medio-ambiente/calculo-emisiones/Ecogan_Informac_default.aspx

> In the same watering system with the same machine I provide as much water as I want them to have that day. In winter I supply water twice and in summer three times. The dose is based on weight, so they get the amount of water that… I set the number of waterings, but the amount they get is based on weight. At 12 I have automated one watering, at 5 in the afternoon another, and at 8 in the evening another in the summer. And in winter… How much does the pig weigh, 100 kg? Well, the machine calculates how many liters of water it has to pour out for the number of animals in that valve. They call it a valve because it is an irrigation valve that opens to let the food drop down. Let's say the valve is in the center of the shed and a stable on each side. So, a certain number of liters of water that corresponds to their weight is released, and the food is the same. I've got a computer over there. (E4, Core manager, G2 factory farm, December 02, 2023)

Feeding is a specific item on the list of Available Technical Improvements, a roadmap for digitalization in livestock farming (MAPA, 2017). Its objectives include reducing input consumption given the progressively rising cost of raw materials, reducing polluting emissions and improving animal welfare conditions. However, despite the opportunities that digitalization offers on paper, in real life environmental erosion and excessive resource consumption continue unabated. In this sense, digitalization can be seen as a false illusion of control over resources and the pollutants generated, where thorough recording of raw materials does not translate into any real reduction in the extraction, exploitation, and appropriation of human and non-human resources (Moore, 2020). On the contrary, it fuels the intensification of production, worsens job insecurity, and reduces animals' lives to mere production units, where every measure to improve animal welfare actually serves to accelerate their metabolism:

> Everything we do is to help the animal gain weight, be comfortable, and eat, because animals gain weight by eating. So, in short, we're going to create the right conditions for the animals to be comfortable, and comfortable animals eat and get fat, right? That's the basis for all this [digital] equipment and such. (E12, manager in a company providing digitalization services in livestock farming, 01/03/2024)

Conclusions

The industrial organization of pork production entails an extractive logic of nature and labor. The resulting health and environmental risks are addressed by companies and farms through what Alexander (2000) calls "an expectation of salvation." Digitalization is framed as a means of risk control that simultaneously increases the economic efficiency of the agro-industrial model. In this way, technology is expected to offset - and even strengthen- the trend towards production centralization and large-scale pork production, along with its inherent extractive logic.

Many analysts have highlighted the "illusion of control" of the agro-industrial model in terms of minimizing risk (Stuart, 2008; Worosz et al., 2008). The claim of compatibility between the continuity of extractivism and the control of its risks is paradoxical. Even more so as the very claim of control via technology and other costly investments encourages the concentration and centralization in factory farming and, therefore, enhances the extractivist logic. Numerous contradictions persist, despite the promises of digitalization to control viral spread: new infections and viruses emerge, water sources are overexploited and slurry contamination continues.

Perhaps the clearest indicator of the illusion of digital farming control, in its attempt to isolate the farms from the outside world, is the persistence of African swine fever (EFSA, 2024). Ultimately, the dream of total control over industrial pork production by means of developing digital monitoring devices remains a corporate fantasy, a technocratic narrative that masks the impossibility of controlling nature, which is mistakenly understood as external to human activity.

References

Alexander, J. C. (2000). *Sociología cultural.* Formas de clasificación en las sociedades complejas, Anthropos, Barcelona.

Barry, J. (2012). Bio-fuelling the hummer? Transdisciplinary thoughts on technooptimism and innovation in the transition from unsustainability. In E. Byrne, G. Mullally, & C. Sage (Eds.), *Transdisciplinary perspectives on transitions to sustainability*. Routledge.

Cattaneo, C. A., & Bocchicchion, A. M. (2019). Dinámica socio-organizacional en el sistema agroalimentario. *Revista Mexicana de Sociología, 81*(1), 7–35.

Coordinación de organizaciones de Agricultores y Ganaderos [COAG]. (2019). *La "uberización" del campo español. Estudio sobre la evolución del modelo social y profesional de agricultura,* https://coag.chil.me/post/la-uberizacion-del-campo-espanol-286547

EFSA. (2024). Epidemiological analysis of African swine fever in the European Union during 2023. https://doi.org/10.2903/j.efsa.2024.8809

Goodman. (2023). *Transforming agriculture and foodways. The digital-molecular convergence.* Bristol University Press.

Harvey, D. (1990). *The condition of postmodernity: An enquiry into the origins of cultural change.* Blackwell.

Hoste, R. (2020, September 13). España: líder en costes de producción en producción porcina en Europa. *Euroganadería.* http://www.euroganaderia.eu/ganaderia/reportajes/espana-lider-en-costes-de-produccion-en-produccion-porcina-en-europa_56_6_1161_0_1_in.html

McMichael, P. (2016). *Regímenes alimentarios y cuestiones agrarias.* Icaria editorial.

Ministerio de Agricultura, Pesca y Alimentación [MAPA]. (2017). *Guía de mejores técnicas disponibles en el sector porcino.* https://www.mapa.gob.es/es/ganaderia/publicaciones/GuiaMTDsSectorPorcino_tcm30-105316.pdf

Ministerio de Agricultura, Pesca y Alimentación [MAPA]. (2023). *El sector de la carne de cerdo en cifras.* Subdirección General de Producciones Ganaderas y Cinegéticas.

Moore, J. W. (2020). *El Capitalismo en la Trama de la vida: Ecología y acumulación de Capital.* Traficantes de sueños.

Pedreño, A., Giménez, M., & Ramírez, A. J. (2021). Cerdos, acumulación y producción de naturaleza barata. *Relaciones Internacionales, 47,* 143–162.

Real Decreto 206/2020. (2020). de 11 de febrero, por el que se establecen normas básicas de ordenación de las granjas porcinas intensivas, y se modifica la normativa básica de ordenación de las explotaciones de ganado porcino extensivo. *Boletín Oficial del Estado,* 38. De 13 de febrero de 2020. https://www.boe.es/eli/es/rd/2020/02/11/306/con

Segura, P., Zapata, M., & Ibáñez, J. A. (1991). *Porcino en Lorca.* Cámara Oficial de Comercio e Industria.

Soldevila, V. (2006). El sector porcino en el estado español a principios del siglo XXI. En Miren Etxezarreta (Coord.), *La agricultura española en la era de la globalización.* Servicio de Publicaciones del ministerio de Agricultura, Pesca y Alimentación.

Stuart, D. (2008). The illusion of control: industrialized agriculture, nature, and food safety. *Agriculture and Human Values, 25*(2), 177–181. https://doi.org/10.1007/s10460-008-9130-4

Worosz, M. R., Knight, A. J., & Harris, C. K. (2008). Resiliencia en la industria de la carne roja de EE.UU.: el papel de la política de seguridad alimentaria. *Agriculture and Human Values, 25*(2), 187–191. https://doi.org/10.1007/s10460-008-9127-z

Digitalisation for Green Agrarian Extractivism in Mediterranean Europe: Dynamics of Change in the Agrifood Enclave of Foggia, in Apulia (Italy)

Carmelo Buscema, Francesco Saverio Caruso, and Alessandra Corrado

Introduction

This chapter aims to investigate how digitalisation strategies serve to 'green' agrarian extractivism and to explore to what extent techno-fixes reproduce a fundamentally unsustainable agrifood system. Agrifood digitalisation is promoted as an instrument to achieve food security, higher levels of productivity, and good practices to cope with climate change. However, from a critical perspective, it appears as a leverage that supports

C. Buscema • F. S. Caruso • A. Corrado (✉)
Department of Political and Social Sciences, University of Calabria, Rende (Cosenza), Italy
e-mail: carmelo.buscema@unical.it; francesco.caruso74@unical.it; alessandra.corrado@unical.it

S. Narotzky et al. (eds.), *Agricultural Extractivism in the Mediterranean Region*,
https://doi.org/10.1007/978-3-032-05599-6_9

capital accumulation and fosters the interests of corporations and particular social forces in the context of the contemporary crisis of capitalism.

A 2020 study by the European Commission (EC) titled 'Farmers of the future' (Krzysztofowicz et al., 2020) already explored the 2040 landscape of farmers dealing with new challenges—that is climate change, technological change and food demand—and the policy implications in the medium- and long-term future. The report asked 'Who will be the next generation of EU agriculture, the farmers of the future?', somewhat downplaying the peasant-style way of farming—characterised by low levels of external inputs, debt, and fossil energy use, and relying on a self-managed resource base—and instead foregrounding new high-tech solutions. In doing so, it prefigured a technology-based strategy embedded in the political reforms promoted in the subsequent years, with an unexpected acceleration and vigour, but justified by multiple crises, from the COVID-19 emergency to climate change and the Ukraine-Russia conflict.

In particular, as part of the Green Deal for the ecological transition programme launched by the European Commission in 2019, the reform of the Common Agricultural Policy (CAP) and the European Union (EU) Next Generation Plan have respectively influenced context and priorities, particularly regarding food security and the resilience of the agricultural sector, while also supporting post-COVID-19 pandemic economy recovery. Both policies aim to make the EU more digital and green. The idea of 'twin transitions' is central to the European Green Deal: 'The term "twin transitions" refers not only to two concurrent transformational trends (the green and digital transitions); the term also refers to uniting the two transitions' (Muench et al., 2022).

The European Commission aims to make the green and digital transitions mutually reinforce each other. On the one hand, the digital transition focuses on the use of digital technologies to increase productivity and economic growth; on the other hand, the green transition, originally led by governments, aims to decarbonise economic activity and mitigate climate change. The digital transition is supposed to contribute to the green transition in two main ways. Firstly, through its ability to generate and manage information flows and optimise processes. Secondly, through its contribution to the dematerialisation of goods (Muñoz de Bustillo Llorente, 2024).

However, framed as a lever for the sustainable development of agricultural productivity and climate change mitigation, agrifood digitalisation, as a way of mobilising human and non-human nature, labour, land, financial capital and knowledge for intensive production, results in a model of 'agrarian extractivism' that is legitimised by new green efforts (Alonso-Fradejas, 2021; McKay, 2017; Veltmeyer & Ezquerro-Cañete, 2023; Ye et al., 2019). Thus, in green agrarian extractivism, conventional resource extraction by intensive agricultural production coexists with or is converted into 'green' extraction through 'sustainability-oriented' practices and production (Alonso-Fradejas et al., 2024; Dunlap et al., 2024).

Grounded in the province of Foggia (Apulia), this chapter draws on critical political economy, ecology and sociology to examine how digitalisation-driven transformations in industrial tomato production affect labour and socio-ecological reproduction. The case of processing tomato production is emblematic of the pitfalls and uncertainties associated with digitalisation. Since the 1980s, the province of Foggia, specifically the Capitanata area,[1] has experienced intensive development of tomato production, through monocultural specialisation, structural reorganisation (i.e. scale expansion, indirect production, etc.), the exploitation of natural resources—land and water—and of labour, gradually racialised by the recruitment of migrant workers. The exploitation of human and non-human nature, widely documented and increasingly addressed through policies and social mobilisations (Tagliacozzo et al., 2021), has occurred within a productive context that has recently shifted towards technology and digitalisation.

This case study illustrates how 'extractivism is no longer limited to the periphery' (Ye et al., 2019), and neither is green agro-extractivism limited to flex crops and commodities[2]—which have underpinned debates on 'sustainable agricultural intensification', 'green commodities' and 'climate-smart agriculture', despite their dependence on fossil fuels and capital-intensive external inputs.

Against this background, the chapter addresses two main questions. Firstly, how is digitalisation emerging as a strategic response to labour conflicts and resistance, and what are its implications in the restructuring of power relations, the reconfiguration of productive knowledge, and the stratification of labour classes? Secondly, to what extent is digitalisation redefining the dynamics of accumulation, control and artificialisation of

[1] 'Capitanata' is a toponym indicating a geographical and cultural region in northern Apulia, comprising the Tavoliere, the Gargano and the Subappennino Dauno.

[2] Flex crops and commodities are those with multiple uses (food, feed, fuel, industrial material) that can be, or are thought to be, flexibly inter-changed (Borras et al., 2015).

nature, and what risks does it pose for the socio-ecological sustainability and survival of 'material civilisations' (Arrighi, 2014)?

This chapter critically examines the digital and green transition through the case of agricultural transformations in the Foggia agrifood enclave, based on fieldwork since 2018 until the time of writing. The focus has been on migrant workers' conditions and mobilisation, with particular attention to agrifood digitalisation in 2024–2025. The analysis combines a literature review, empirical data from participant observation, 23 interviews with key stakeholders, and informal conversations with workers.[3]

The initial sections reconstruct, adopting a historical perspective, the process of organisation of the Mediterranean agro-food enclave of Foggia according to the model of agrarian extractivism, highlighting its interconnected characteristics: '(1) large volumes of extracted materials for export with little or no processing; (2) value chain concentration and sectoral disarticulation; (3) high intensity of environmental degradation; (4) deterioration of labor opportunities and/or conditions' (McKay, 2017, p. 202). Next, the transformation-for-sustainability of the extractivist model by means of digitalisation is analysed. Final conclusions follow.

The Organisation of an Intensive Agrifood Enclave

Apulia, a region located in southern Italy, has a strong agricultural vocation due to a morphological configuration which is predominantly flat. It is the Italian region with the highest incidence of Utilised Agricultural Area (UAA), about 65.4% (1,285,274 hectares, corresponding to over 10% of the entire Italian UAA), whereas at the national level the average is just over 40% (ISTAT, 2023). Cereals cover over half of the cultivated land in Apulia (675,739 ha), followed by olives (366,896 ha, nearly 30% of regional UAA and the largest in Italy) and vines (92,038 ha, over half of the country's table grapes).

The Tavoliere, Italy's second-largest plain, hosts intensive agriculture and has seen a shift from wheat to more profitable fruit and vegetable crops in recent decades. Foggia alone accounts for over 50,000 hectares—contributing to about 30% of Italy's open-field production of vegetables like asparagus, cabbage, aubergines, artichokes, lettuce, fennel, broccoli

[3]The chapter is based on the results of the National Research Project (PNRR) 'DiJjust - Digital Food and Just Transition. Sustainability and Labour in Agriculture 4.0' (N. P2022X3MWR, CUP 53D23006850001).

and peppers. These numbers illustrate how what was once called the 'granary of Italy', is increasingly becoming the 'orchard' of Italy.

The rural landscape of the area—once a humid, marshy and unhealthy plain of over 400,000 hectares where mosquitoes and malaria thrived—was profoundly transformed as early as the 1920s with the reclamation of vast swampy areas for agricultural exploitation,[4] deforestation, and intensive urban planning, under fascist rhetoric (i.e. the 'battle of wheat' launched in 1925 for Italy's cereal self-sufficiency). In the post-World War 2 period, with the establishment in 1965 of the Capitanata Consortium of water management (*Consorzio di bonifica*), a powerful system of reclamation and hydraulic infrastructure was developed in the area, based mainly on the channelling of the Ofanto and Fortore rivers and the support of four artificial basins (Occhito, Capacciotti, Capaccio and San Pietro).

After the unification of Italy, land ownership in Apulia became concentrated in the hands of a few large landowners, strengthening the *latifundia* instead of favouring small peasant properties. At the turn of the twentieth century, agrarian capitalism was based on large monocultures and mechanisation, with continuous conflicts between landowners and masses of landless labourers. This gave rise to trade union movements and struggles for the redistribution of land that culminated between 1946 and 1952. However, concessions to labourers did not resolve the agrarian question. Peasants were employed mainly as day labourers, for the peak work of weeding and harvesting, with rare cases of precarious co-tenancy and small-rental relationships.

This important period of strikes and land occupations throughout southern Italy led to the Land Reform (Law No. 841/1950)[5] that gave birth to a small agricultural bourgeoisie—through the bestowal of estates of six-to-ten hectares. In the plain nearly 17,000 people, 60% of the entire province, applied for land. About 6000 farms were established, for a total of about 30,000 hectares. But, as elsewhere, land reform was not sufficient to establish sustainable farms and within a few decades many farms went bankrupt.

[4] Land Reclamation advanced during the 1920s. In 1927 the first private land reclamation consortia (by the initiative of landowning families) were set up, later unified in the Consorzio Generale di Bonifica della Capitanata, which, together with the Opera Nazionale Combattenti, contributed profoundly to the rural transformation of the Tavoliere.

[5] Labour and trade union mobilisations also demanded the 'imposition of a labour quota', obtaining a law (in force for around ten years) according to which landowners were obliged to hire a certain amount of workers based on the amount of land they owned.

Over the past 30 years, farmers in Foggia have increasingly invested in tomato production, a symbol of the 'Mediterranean diet' and 'Italian cuisine' or 'Made in Italy' (Perrotta, 2016). While production has expanded in northern regions like Emilia Romagna, the South remains the leading area in term of tomatoes for processing produced. However,[6] environmental degradation and criminal activity have devalued local agriculture, especially in areas around Naples (like the '*Terra dei fuochi*'), leading markets there to label products as 'non-local' to reassure consumers. In addition, in the mid-1990s, a socio-ecological and phytosanitary crisis led to the decline of tomato production in the Naples area, prompting the processing industries to move to Capitanata, an ideal area due to its climate, soil and lower anthropic pressure. This triggered the "red gold rush": thousands of farmers invested in tomato cultivation, attracted by high yields and rising prices. The move was also encouraged by CAP incentives, varietal innovation and the creation of irrigation infrastructure.

As a result, Campania's utilised agricultural area has shrunk by over 30%, while in Apulia it has remained stable at over 1.5 million hectares, with a third of it in Foggia. Here, land concentration has increased, with farms averaging 12 hectares—double the regional average but still small compared to national and international tomato farms. Most remain family run, relying on external labour for planting and harvesting.

The Tavoliere has thus progressively structured itself as the main pole of 'Made in Italy' tomato sauce, with more than 15 thousand hectares dedicated and more than 3500 producers cultivating around 20 million quintals of product, that is, around 40% of the 50 million quintals of total production, mostly of long tomatoes (ISTAT, 2023).

Tomato production is concentrated in the Foggia area, while processing takes place mainly between Salerno and Naples, where 70% of the Italian plants are located. Foggia is also home to Europe's largest processing plant, Princes Industrie Alimentari (PIA) (now part of Newlat Food). The proximity to the fields is strategic, as the tomatoes have to be processed fresh to guarantee quality (interview to PIA CEO, 2021). About 95% of production, including own brands and private labels, is exported.

[6] The physical and economic size of farms in the South in smaller than in the North, which does not allow for economies of scale: compared to the North, there is greater use of subcontracting, a greater need for man-hours, higher costs for certain production factors (seedlings, water, pest control), and less mechanisation; In addition, there is greater recourse to land rental and the plots are distant from each other (CREA, 2023).

Italy is the world leader in exports of canned tomatoes, with more than 2.2 million tonnes exported in 2024, worth €3 billion (ANICAV, 2025).[7]

However, while the business flourished for a few years, several production problems soon began to arise. Firstly, tomato production requires a large mass of labour concentrated in the short summer harvesting phase. Local labour has been particularly reluctant to do this type of work (unlike olive and viticulture), characterised by particularly harsh conditions, heavy tasks requiring physical strength and endurance, harvesting operations concentrated during a limited period in high temperatures, low wages (often piecework), and the irregularity of employment arrangements and intermediation. This led to the recruitment of a considerable number of workers from Eastern Europe, reaching 16,000 in 2012. Having gained full freedom of movement with the eastward enlargement of the European Union, Eastern European workers began to seek better employment elsewhere. In 2024, less than 4000 employment contracts were signed with EU workers in the province of Foggia. In contrast, the number of African workers doubled over the same period (2012–2024) from 4000 to 8000 contracts. It is worth noting that official statistics do not take into account the significant share of the black and grey labour market (Caruso et al., 2024).

Over the years, the temporary presence and lack of inclusion policies for seasonal workers in the region have led to the organisation of self-managed housing solutions, such as isolated 'rural ghettos'. During the summer months, these host several thousand labourers in degrading and precarious conditions, especially the most vulnerable migrants who encounter major obstacles to finding work and accommodation in other territorial and productive contexts in Italy. This is the case of applicants for international protection, who live in the limbo of administrative semi-invisibility (lacking residence, a bank account, a general practitioner, a postal address, etc.) and who find better chances to reproduce making a living in the Foggia countryside. These dynamics have produced the 'refugeeisation of agricultural labour' (Dines & Rigo, 2015).

In addition to this situation, we have to mention the severe exploitation of another indispensable labour segment, consisting of about a thousand

[7] Over 80% of the tomatoes processed in Southern Italy are used for private labels of food retail chains, which leverage their power to undermine established brands and boost their own products. The power imbalance in the tomato supply chain especially harms food manufacturers, small firms and farm workers downstream (D'Onofrio, 2024).

truck drivers, who move millions of quintals of tomatoes from Apulia to Campania, where almost 80% of Foggia's tomatoes are processed. The long distance between production and processing sites causes significant environmental impact, especially in summer, with around 50,000 lorries—each carrying 270 quintals—clogging the Naples-Bari motorway. Drivers, earning little, often repeat the 400 km round trip in a day, facing exhausting conditions and health risks during long waits to unload.

The intensive exploitation of labour is also compounded by the intensive exploitation of land. The search for profit maximisation has pushed farmers to force crop cycles and rotations. The tomato crop requires a strong level of rotation that should inhibit its cultivation on the same land for at least two or three years, in order to reduce the pathogens in the soil and prevent disease outbreaks. The extractivist approach has resulted in the unregulated use of plant growth regulators, such as gibberellic acid, aimed at accelerating tomato breeding. Crop cycles are thus progressively shortened and the resulting effects are countered through the intensification of chemicals inputs and plant protection products in an attempt to counter the exponential growth of diseases—particularly downy mildew, powdery mildew and ToBrfv (tomato brown rugose fruit virus)—as well as the spread of pests such as orobanchids, which are reducing yields and undermining profitability.

The transnational seed companies' strategy to cope with these diseases by experimenting with more resistant hybrid seeds has led to increased costs for farmers. However, these costs have not always translated into improved results in the field, partly because mass planting has been favoured over the traditional density of 30,000 seedlings per hectare (Farmer, interview no. 2, 2025).

A further critical aspect of extractive tomato production in the Tavoliere is the unsustainable water impact (Ciervo, 2013), with the long tomato requiring an estimated 7000/8000 cubic metres of water per hectare (OP & ANICAV, 2020). In recent years climate change and severe drought have depleted water resources in the area. The water flow of the Occhitto Dam, the largest reservoir, has dropped from 180 million in 2019 to 120 million in 2023, reaching the current threshold of 34 million. This is almost ten million below the so-called dead volume (i.e. the minimum holding threshold between the level of exhaustion discharge and the foundation level), below which other reservoirs are also beginning to fall.

Local institutions, faced with this ecological crisis, could do nothing but issue on August 13, 2024, in the midst of the tomato campaign, a historic ordinance that warned farmers that 'The distribution of water for irrigation use in the areas of the Capitanata Consortium ends today. The available water resource will be used for priority use drinking purposes' (Consorzio di Bonifica della Capitanata, 2024).

APO Foggia, the largest consortium of tomato producers representing 40 cooperatives, 900 producer members and 6000 hectares of cultivated land, had warned its members to anticipate the planting in light of a scenario of water collapse. However, many producers were caught off guard and financially devastated by the early shutdown of water access, which forced them to rely on tankers to save their crops, triggering a sharp rise in production costs.

Therefore, the social and environmental unsustainability of tomatoes is now also associated with economic unsustainability: although in recent years the price has settled at around 16–20 cents per kilo, more than double its value compared to a decade ago, it is no longer sufficient to cover the significant increase in both direct and indirect costs.

The extractivist process seems to have reached its limits, as new globally demanded crops are introduced—assuming Foggia's agriculture manages to endure the current crisis.

Green Extractivism by Digitalisation

Numerous tragic events (murders, deaths from car accidents during transport to the camps in overcrowded vehicles, bicycle accidents on poorly lit roads, fires in rural ghettos, exploitative working conditions), documented cases of exploitation, violence, trafficking and irregular gangmastering of migrant workers, policy failures (Caruso, 2022; Fanizza, 2013; MSF, 2005; Perrotta, 2014), as well as mobilisations by migrant workers (Corrado et al., 2023; Perrotta & Sacchetto, 2014), have gradually attracted the attention of the international media, human rights organisations, trade unions and critical consumers.

Under such pressure, national policies addressing undeclared agricultural work have evolved, notably with Law No. 199/2016, which redefined the crime of *caporalato* to also impose sanctions on employers exploiting workers. The law also extended protections to victims. However, despite some progress in tackling irregular migrant labour, these measures have shown limited effectiveness, as deeper structural changes in supply chain organisation and local policies are still lacking (Palumbo, 2024).

It is in this context that the debate on digitalisation emerged in Italy, animated first—in the wake of the COVID-19 emergency—by employers' alarm at the lack of workers and then by the EU Recovery Plan and CAP subsidies, aimed at a greener, more digital and more resilient Europe.

Since 2018, various digital platforms have been launched to improve labour conditions and match supply and demand, especially during COVID-19. However, they had limited success due to technical inefficiencies, employers' resistance to perceived bureaucracy and oversight, and the low digital literacy of many migrant workers. The digitalisation process involving local farmers is of an exogenous nature, as it emerges on the basis of external solicitations, from public and private actors, outside the farmers' cultural and professional world of reference. The various private and public initiatives promoting the digitalisation process in agriculture identified at different times in our research help us to understand the role of different actors, changes over time, and the accompanying rhetoric, mainly reproduced by institutions, researchers and trade organisations.

In 2019, a collaboration between PIA and Coldiretti in Foggia was born, following pressure from retail chains in Northern Europe concerned about working conditions in the tomato supply chain. These companies financed the Ethical Trade Initiative (ETI) intervention in Foggia, which involved trade unions, NGOs, local and national institutions and companies, to improve conditions for workers and producers through the negotiation of standards (IEH et al., 2015).

> The management principles of the Mitsubishi Corporation place great emphasis on respect for human rights and the centrality of people in all business processes', PIA CEO explained. A number of investments were therefore geared towards ensuring ethical compliance. In fact, PIA is asked by its international customers not only to 'guarantee the quality of its products', but also to 'ensure respect for human rights throughout the supply chain. (interview with PIA CEO, 2021)

Princes has an agronomy department, which during the tomato season is responsible for supporting all farmers in their agronomic practices, but also for ensuring 'ethical compliance' throughout the supply chain and at all stages of the tomato growing process. The required practices consist of the adoption of a series of quality certifications. ETI membership by PIA has meant that processed tomatoes are produced by companies that adopt the Global GAP/GRASP (Global Good Agricultural Practice/Risk

Assessment on Social Practice) and SA8000 certifications issued by SAI (Social Accountability International). As of 2020, producer-suppliers of PIA were also asked to register with the Quality Agricultural Labour Network established by Law n. 199/2016.

PIA and Coldiretti Foggia jointly promoted a supply chain agreement aimed to ensure all-round sustainability of the entire tomato industry chain. The agreement consisted of specific elements: (a) the shortening of the supply chain to counter irregular gangmastering; (b) the introduction of mechanised harvesting and a specialised and stable workforce; (c) quality control throughout the supply chain through technical assistance, digitalised input control and a blockchain system; (d) the definition of a fair price (around €120 per tonne).

Contracts with these conditions were signed with nine producer organisations (POs),[8] which in turn bring together 35 cooperatives, involving some 300–350 growers.

The production protocol imposed on farmers includes stringent and costly requirements that are not always perceived as useful. However, while dealing with the supermarket trade—often mediated by commercial intermediaries—tends to squeeze profit margins, exports, particularly to the UK under the Napolina brand, offered greater valorisation (farmer-processor, interview n. 3, 2025).

In addition to complying with quality certifications, suppliers were asked to accept other contractual obligations regarding sustainability. Princes asked for the adoption of digital technologies aimed to save irrigation water and nitrogen use in the open air tomato fields (namely a Decision Support System [DSS] called GesCoN/Ecofert), and to reduce the use of pesticides by a web- and cloud-based system. 'Traps' in the fields provide an indication of which specific pest is attacking the crops, so that the farmer is advised on when to irrigate and which type of pesticide to apply (Trapview by the Slovenian enterprise EFOS).

PIA CEO interviewed by the authors in 2021 emphasised how this change had 'important results in terms of quality', as 85% of PIA's production in the last years has been completely free of chemical residues, 'despite

[8] Producer organisations and their associations help farmers to reduce costs, improve quality management, and collaborate in the processing and marketing of products. They are supported by the CAP to strengthen the collective bargaining power of agricultural producers.

the fact that the law allows a minimum percentage of chemical residues in food'.

Farmers were required to use two digital platforms: the PIA platform and Coldiretti's blockchain platform Tokenfarm. This added an 'extra job' to their workload, often accompanied by significant difficulties in using these platforms, such as lack of time and limited digital skills (interview with Coldiretti Foggia director, 2021).[9]

The Coldiretti start-up Tokenfarm aimed to use blockchain technology to ensure transparency and fairness in supply contracts across the tomato chain. While the system promised trust through data accessibility and traceability, its environmental impact (due to its high energy use) and limited adoption by farmers three years after launch, have raised concerns about its effectiveness (interview with Coldiretti Foggia director, 2025).

Farm Selection and Resistance

The promotion of digitalisation in the fields—through partnerships involving first the University of Foggia and then the Catholic University of Milan, with the spin-off Hort@—was met with a certain 'resistance' from companies to adopt and use digital technologies, according to the Hort@ consultant interviewed. The consultant discussed differences in the approach to sustainability in agriculture. The collaboration between Hort@ and PIA is based on a decision support system, which aims to improve sustainability and overall performance for agribusinesses. However, the implementation of such systems is complex and takes years to achieve significant results. The consultant emphasised that in intensive growing contexts such as tomatoes, characterised by fast and dynamic production cycles, the integration of new processes is particularly challenging. Farmers tend to focus on technical efficiency and improving quality, rather than on sustainability considerations. Indeed, in the past, 'the concept of sustainability was perceived more as a policy buzzword rather than a concrete goal'. Only in recent years has sustainability represented a significant challenge for agriculture, mainly due to external pressure from large companies such as PIA and growing consumer expectations for reduced

[9] Coldiretti requires its members to use the Digital Farm Notebook, a tool that allows integrated and real-time management of all farm activities, from treatments to irrigation, from fertilisation to accounting. Thanks to the integration with the Demetra portal, it also offers data on crop health, weather, soil fertility and CAP subsidies.

environmental impact. However, the adoption of sustainable practices is often fragmented and varies greatly between different production units.

A Hort@ consultant highlighted how Decision Support Systems (DSS) could assist the management of variability in fields, suggesting, for example, the use of intelligent machines for targeted operations. However, the adoption of such tools depends on 'farmers' propensity for digitalisation', rather than on the direct influence of DSS. For crops such as tomatoes, DSS provides useful guidance for pest management and other agricultural practices, but does not directly determine mechanisation choices. 'It is an indirect relationship, where the digitally oriented farmer is often also more predisposed to automation'.

The consultant contrasted the sustainability orientation of big companies, based on implementing procedures, adhering to standards, optimising control systems, and customer demands, with that of the smaller farmer who is more concerned with yields and costs.

Farmers' national associations have created their own digital platforms, spin-offs involving different service companies, in collaboration with software houses—high-tech enterprises in articulated partnerships. These platforms—Demetra of Coldiretti, Hubfarm of Confagricoltura—aim to improve the organisation of agricultural campaigns, connect machinery, equipment, sensors in the field and integrate information from satellites or drones. However, a limited number of farmers are adopting these new technologies promoted by their trade associations in Foggia.

Analysis of data from the last agricultural census shows the low level of informatisation of farms nationally. This is attributed to age and educational level, but also to a lack of connectivity in the countryside. However, it is also important to read this figure in the light of the territorial and dimensional divide: one-third of farms in northern Italy are computerised, compared to 6.7% in the south; among the large farms with more than 50 hectares, about half are already computerised, while among the 240,000 very small farms only around 15,000 use computer or digital support. The rate of informatisation is 8% for farms with one employee, while it is 78% for farms with more than ten agricultural work units (AWUs) (Gnesi, 2022).

According to the Director of the local branch of the Farmers' National Association Confagricoltura 'only a few entrepreneurs considered as "enlightened" have been able to approach the new technologies, but in reality we are still in the Middle Ages, (...) halfway between North Africa and Europe' (interview with Confagricoltura Foggia Director, 2024).

The Director of the local branch of Coldiretti observed that even when farmers purchase digitally equipped machinery and tools thanks to advantageous funding, 'they do not use them or do not use them properly because they lack the appropriate skills or knowledge'. This is also why the collaboration with PIA has encountered limitations in implementation, according to him. Only a few farms seem to have the capabilities to be innovative through digitalisation (interview with Coldiretti Foggia director, 2025).

On the basis of this evidence, a solution is being sought by Coldiretti through a national project financed by the NRRP: Digital Agriculture Hub. This innovation hub aims to enhance the digitalisation of the agricultural sector across the country through three main phases: assessment of the digital maturity of farms; orientation and training of farms; technology transfer in farms. The project aims to provide services for small farms with up to 49 employees, with a target of 8900 farms.

EU funds earmarked for digitalisation in agriculture and provided for in the Regional Development Plan have in a way 'imposed' the purchase of licenses, applications and machinery on farmers. These policies reward the companies with the greatest extension, further supporting the dichotomisation between small and large companies, as in the case of the selection by private initiatives, that is, the Digital Agriculture Hub.

Public subsidies for digital agriculture are a strong incentive for modernisation, granting up to €800 per hectare to some applications. Today's subsidies are much higher than in the past (where they were €80–120/hectare), making the adoption of the most advanced tools more convenient. However, only a few manage to meet all the requirements to apply (farmer and subcontractor, 400 ha, interview n. 4, 2025).

Some farmers access the subsidy through subterfuge, in complicity with business consultants and software houses backdating the diachronic georeferencing featured as a prerequisite. They introduce the technologies in order to benefit from the subsidies and renew their machinery but—as other actors said—do not know how to use them (consultant, interview n. 1, 2024).

Several interviews confirmed the impact of subsidies in driving the adoption of digital technologies, but also the pressures from industry and other actors downstream in the supply chain. 'The incentives prompted many companies to overcome resistance to change and adopt machinery such as mechanised tomato harvesting, also at the request of customers such as the La Doria group'.[10] (farmer and subcontractor, Interview n. 4, 2025).

Labor and Nature Waste

The process of digitisation of agriculture also affects work, transforming employment conditions, work processes, and relationships with human and non-human nature. The rhythms and logics of cost-cutting and standardisation produce superfluous human labour (as an alternative to exploitation) and agrifood products abandoned in the fields or damaged without care. Human and non-human nature turn into waste (Armiero, 2021).

Mechanical harvesting, pushed by the big companies, is justified with several arguments. Mechanised tomato harvesting can use optical devices (colorimeters) based on digital technology to analyse and sort tomatoes. These devices, such as optical sorters, employ sensors and cameras to assess colour, shape, the presence of defects and foreign objects, and discard those that do not conform.

PIA demanded mechanised harvesting from producers to address the problem of labour exploitation: mechanised harvesting 'does not eliminate the problem of exploitation, but reduces it, as in an average field (5 hectares), mechanised harvesting normally requires six people working in one day, while manual harvesting requires 50–60' (interview to PIA CEO, 2021).

Mechanical harvesting (often subcontracted) is also a response to labour resistance through mobility or conflict over labour remuneration (often treated as 'labour shortage').

> [It] proves to be more advantageous, despite a slight loss in product consistency. The costs of manual harvesting become unsustainable due to labour shortages. (…) a harvesting machine can cover one hectare for €600, while manually the cost would be much higher. (…) sometimes the service [by subcontractors] also includes loading onto the truck and the operators, depending on the agreement. (Farmer, interview n. 5, 2025)

> The machine allows harvesting to be completed quickly, avoiding problems of absenteeism or worker exhaustion. Some farmers are willing to pay a little extra for manual harvesting, which is considered more delicate, but much

[10] La Doria went from being a small tomato processing factory to a global leader with a turnover above €1 billion.

> depends on the farmer's ability to manage the field, the variety of tomato and the time of harvest. (Farmer-subcontractor, interview n. 4, 2025)

On the labour side, on the other hand, a certain trend has emerged that places it in conflict and direct competition with digitalisation and, more generally, with the mechanisation of harvesting. Mechanical harvesting is opposed by farmworkers made redundant. Farmworkers see not only their employment opportunities but also their economic conditions compromised.

This issue was discussed during an informal collective interview with a group of four unionised migrant labourers living in the Torretta Antonacci informal camp, with whom one of the authors shared a long car journey.

> The machine is much cheaper than manual labour, due to the smaller number of workers employed and the speed of operations. Using the machine requires only six or seven people to manage the entire process: two forklift drivers to transport the boxes, two workers to control the tomatoes, and a small number of workers to sort the product and clean it. This process, although mechanised, does not require highly specialised skills and can be performed by anyone, unlike manual work which requires more experience to avoid damaging the tomatoes.
>
> Tomato harvesting machines are generally not bought by the farms, but rented. The rental cost depends on the quantity of tomatoes harvested: for each truck, the rental price of the machine is around €450. The modern machines are faster and can harvest tomatoes for up to 12 trucks per day, while the older models reach a maximum of five or six trucks per day.
>
> In manual work, a team of about 10 people can harvest a truckload of tomatoes in about one and a half hours. In contrast, the machine allows the same amount to be harvested with fewer people and in less time. The machine allows faster harvesting, making the work more comfortable, less tiring for the workers; whereas manual work is physically more demanding, causing fatigue for those doing it. Finally, regarding compensation, workers with the machine generally earn between €80 and €100 per day, while those who work manually can earn a similar amount, despite the physical work being more strenuous. In manual harvesting, workers earn around €6 per box. Some pickers manage to complete up to 30 or 40 skips in a morning, thus earning a significant amount. However, this type of work is subject to the uncertainty of daily earnings. (Fieldnotes, November 2024)

Against the risk of reduced job opportunities, threats of 'luddism'—the boycott of machines—and 'rain dances'—on muddy ground, machines cannot operate and harvesting must be done by hand—are to date the only instances of resistance. In the agrifood sector there is still a lack of response on the part of trade unions regarding the relationship between technology and wages. Harvesting machines are seen as tools that risk jeopardising the daily wages that are earned in the tomato summer period through piecework. In addition, the most controversial point remains the misalignment of hours and wages between work teams equipped with mechanical-digital support—whose daily pay falls within the contractual parameters of the general and unskilled agricultural worker, with recurring and abundant overruns of the scheduled working hours—and teams employed in manual harvesting, who manage to obtain even double the pay with half the working hours, albeit with high rates of exploitation.

The application of digital sensors to support mechanised harvesting paradoxically seems to affect the quality of the product too, and runs counter to the very principle of optimisation and efficiency. In fields where harvesting is carried out by means of harvesting machines, food waste is produced, with quintals of tomatoes left to rot in the fields due to lack of perfect sizing and ripening. But, for an interviewed farmer, this also depends on farmers' abilities or skills in customising the optical selectors of the machines.

> Mechanical harvesting is faster and less exposed to labour, but it can cause trauma to the fruit if not well managed. (...) optical selector settings are customized to the quality standards required by the industry. The quality of the harvested tomato is closely linked to the varieties used and the farmer's care in managing the plant. Today, it is possible to harvest a product with up to 90–95% ripe fruit, thus reducing the amount of waste. However, this result depends very much on the farmer's ability to prevent defects such as apical rot or the presence of green fruit, which are often more a psychological perception than a real quantity. Environmental conditions also play a key role in determining the ripeness and quality of the harvest. (Farmer-subcontractor, interview n. 4, 2025)

During one campaign, second-generation American machines with sensors that select good tomatoes was observed in use. A farmer pointed out that while these machines reduce the need for manual labour, they can shift the cost of rejected produce back onto the farmer. It was reiterated that, if well organised, manual harvesting can be more effective (interview n. 6, 2025).

Nature Depletion

The impact on environmental sustainability of these new digital technologies is a controversial issue. This is especially true in the case of water management.

Water availability directly affects irrigation capacity; many farmers, lacking wells or alternative resources, find themselves growing crops (i.e. non irrigated crops) that have lower profit margins. This problem also affects the agricultural calendar: limited irrigation compromises production, especially during crucial periods such as August.

Farmers receive an allocation of water resources from the consortia based on the size of their land and the type of crops they grow. However, wells and other variables introduce 'management complexity', making the current system inflexible. The physical infrastructure of micro-irrigation systems, now a common element in tomato fields, has for years been the indispensable strategy to combat growing drought in the area.

Several DSSs, developed by corporations or supplied by Israeli companies, experimented to develop an electronic irrigation management system, capable of 'collecting' information from the field and processing it according to defined agronomic procedures for determining precisely when to irrigate, and the corresponding volume of water needed for the industrial tomato crop.

Some pilot research projects by public-private partnerships aiming to save water show little or no efficacy (researcher, interview n. 7, 2025).

Difficulties arise from the structural characteristics of most companies: irrigation units are often organised in one-acre blocks; this configuration limits the ability to detect local variability and optimise irrigation in a granular manner. The lack of integration between irrigation automation and information systems is also considered a problem for the efficacy of DSS, not just for water saving but also for enhancing productivity by reducing farmer intervention, especially on large surfaces.

Interviews with experts point to an ongoing process: the attempt to overcome the limitations of manual irrigation management through the introduction of automated and predictive systems. At present, irrigation control requires the farmer to manually input data, an operation that consumes significant time and resources. In response to this critical issue, agronomic research is developing solutions that integrate existing agricultural machinery with automated technologies in order to simplify farmers' work and improve the overall efficiency of water management.

However, the available technologies still have important limitations. Soil moisture sensors provide point data, which often do not reflect the full variability of the field and take time to calibrate. Furthermore, they are unable to anticipate future crop water requirements. To overcome these limitations, mathematical models based on evapotranspiration are being developed, offering more reliable estimates by considering different agronomic and environmental parameters.

The future of irrigation seems to be oriented towards the integration of predictive models and sensors for more dynamic management of water resources. However, this scenario still requires technological advances and, above all, substantial economic investment, which is not accessible to all farmers.

Farmers' practical yet 'old-fashioned' knowledge seems to have definite advantages, especially in the event of unforeseen events. In addition, digital technologies can do nothing in the face of dramatic local drought conditions, as well as the salinisation of water. So, 'the real question is not the renewal of supply chain agreements or the digital transition, but whether there will still be water tomorrow' (Foggia Coldiretti Director, 2024).

> Although automation is perceived as advanced, many large farms still use personnel to manually open and close water valves. This approach, although seeming old-fashioned, offers practical advantages. The farmer operating manually directly monitors the water conditions, whereas full automation may not be flexible enough in case of emergencies, such as a sudden pipe break. When this happens, the farmer must decide whether to skip a sector and compromise the irrigation cycle or stop and reschedule. The idea that automation is always the ideal solution is challenged, as manual decisions, although less efficient, can be more adaptable. (Hort@ consultant, interview n. 2, 2024)

> Intelligent irrigation systems are being experimented with, connected to control units and sensors in the ground, automatically indicating and regulating irrigation shifts, opening or closing valves in the various sectors. These technologies make it possible to optimise the use of water. However, in the event of drought, wells can also suffer flow and quality problems, with salinisation phenomena, especially in the more superficial aquifers. The processing tomato is the horticultural crop most exposed to these difficulties, as it is mainly cultivated in summer and requires large quantities of water. (Farmer-subcontractor, interview n. 4, 2025)

Conclusions

The case of the Foggia agrifood enclave supports the idea that the digitalisation strategy is the lever of a new manifestation of agrarian extractivism through a twin transition, both green and digital.

In order to understand the impact of digitalisation and assess the extractivist intensity of this transformation-for-sustainability, following Alonso-Fradejas (2021), we have tried to consider: the forms of value control and capital flows connected to digitalisation (i.e. value chain dynamics and their impact on investments and capitalisation); employment and labour dynamics (i.e. conflicts, working conditions and characteristics, labour differentiation, the replacement of land workers); social metabolism dynamics (i.e. conditions of reproduction of human labour); political-normative dynamics (i.e. the role of regulation and main narrative).

Digitalisation in agrifood is a strategy to expropriate human and non-human nature and its conditions of reproduction, and to appropriate the value generated in long chains under centralised control. The subsumption of both farmers and labourers within this process occurs through interconnected top-down mechanisms, driven by state and market forces. These processes are framed as sustainability-oriented responses to the current climatic-environmental, economic-financial and geopolitical crisis.

Digitalisation responds to labour conflicts by replacing labourers with machines, embedding and centralising productive knowledge in algorithms and digital tools. This process disempowers agricultural labourers, making them more replaceable and controllable, while deepening labour stratification and shifting wealth and power from labour to capital.

Digitalisation also extends the dominant digital business model into agriculture, enabling upward redistribution of wealth—from producers to industry, digital enterprises and finance—under the flag of 'sustainability'. This involves green technologies, supply chain control and due diligence mechanisms, leading to further artificialisation/dematerialisation of nature and a deeper metabolic rift.

However, digitalisation processes reveal limitations in achieving the anticipated optimisation and efficiency. These limits stem not only from forms of resistance aimed at preserving margins of autonomy, self-control or reproduction—including practices that involve the exploitation or depletion of nature—but also from the inherent constraints of the ecosystem itself. The alternative to the destruction of nature seems to be other than a salvific digitalisation.

References

Alonso-Fradejas, A. (2021). Leaving no one unscathed in sustainability transitions: The life-purging agro-extractivism of corporate renewables. *Journal of Rural Studies, 81*, 127–138. https://doi.org/10.1016/j.jrurstud.2020.10.001

Alonso-Fradejas, A., Ezquerro-Cañete, A., & McKay, B. (2024). La agenda política del "agroextractivismo verde" en América Latina. *Revista de ciencia política (Santiago), 44*(1), 1–21. https://doi.org/10.4067/s0718-090x2024005000101

ANICAV. (2025, aprile 2025). *Nel 2024 export di conserve di pomodoro per circa 3 miliardi di euro (+3,8 % in valore)*. [Press release]. https://anicav.it/wp-content/uploads/2025/04/CS-giornata-made-in-italy_export.pdf.

Armiero, M. (2021). *Wasteocene: Stories from the global dump.* Cambridge University Press.

Arrighi, G. (2014). *Il lungo XX secolo.* Il saggiatore.

Borras, S. M., Franco, J. C., Isakson, S. R., Levidow, L., & Vervest, P. (2015). The rise of flex crops and commodities: Implications for research. *The Journal of Peasant Studies, 43*(1), 93–115. https://doi.org/10.1080/03066150.2015.1036417

Caruso, F. S. (2022). Ghetti rurali e profughizzazione del lavoro bracciantile nell'orto d'Italia. *Mondi migranti, 2*, 37–52. https://doi.org/10.3280/mm2022-002002

Caruso, F. S., Corrado, A., & Macciani, C. (2024). I lavoratori agricoli stranieri dopo la pandemia: dinamiche di cambiamento nel Sud Italia. *SINAPPSI, 1*, 104–118.

Ciervo, M. (2013). Il pomodoro da industria in Italia. Nodi ed effetti territoriali attraverso il caso di studio della Capitanata (Puglia). *Bollettino della Società Geografica Italiana.* Serie XIII, VI, 293–329. https://bsgi.it/index.php/bsgi/article/view/7429

Consorzio per la Bonifica della Capitanata. (2024). Stagione Irrigua: avviso. [Press release] https://www.bonificacapitanata.it/stagione-irrigua-2024-avviso/

Corrado, A., Caruso, F. S., & Macciani, C. (2023). Migrazioni, sindacalismo e autorganizzazione nelle campagne: i braccianti migranti in provincia di Foggia (Italia). *Mondi Migranti, 2*, 133–151. https://doi.org/10.3280/mm2023-002008

CREA. (2023). Analisi economica del processo produttivo del pomodoro da industria. https://anicav.it/wp-content/uploads/2023/11/CREA-Studio.pdf

D'Onofrio, G. (2024). The agro-food industry in Southern Italy and its supply chain: The case of the tomato value chain. *Meridiana, 111*, 103–125.

Dines, N., & Rigo, E. (2015). Postcolonial citizenship and the 'refugeeisation' of the workforce: Migrant agricultural labor in the Italian Mezzogiorno. In S. Ponzanesi & G. Colpani (Eds.), *Postcolonial transitions in Europe: Contexts, practices and politics* (pp. 151–172). Rowman & Littlefield.

Dunlap, A., Verweijen, J., & Tornel, C. (2024). The political ecologies of "green" extractivism (s): An introduction. *Journal of Political Ecology, 31*(1), 436–463. https://doi.org/10.2458/jpe.6131

Fanizza, F. (2013). L'immigrazione nelle aree rurali della Puglia: il caso della Capitanata. In Corrado & Colloca (Eds.), *La Globalizzazione della campagne: Migranti e Società Rurali nel Sud Italia* (pp. 94–112). FrancoAngeli.

Gnesi, C. (2022). Digitalizzazione e innovazione delle aziende agricole italiane. ISTAT. https://www.istat.it/it/files/2022/06/censimento_agricoltura_gnesi.pdf.

IEH, ETI, DIEH. (2015). Due diligence in agricultural supply chains: Counteracting exploitation of migrant workers in Italian tomato production. https://www.ethicaltrade.org/sites/default/files/shared_resources/italian_tomato_production_report.pdf

ISTAT. (2023). L'andamento dell'economia agricola - Anno 2023. https://www.istat.it/comunicato-stampa/andamento-delleconomia-agricola-anno-2023/

Krzysztofowicz, M., Rudkin, J., Winthagen, V., & Bock, A. (2020). *Farmers of the future*. European Union. https://doi.org/10.2760/680650.

McKay, B. M. (2017). Agrarian extractivism in Bolivia. *World Development, 97*, 199–211.

MSF. (2005). *Una Stagione all'Inferno: Rapporto sulle Condizioni degli Immigrati Impiegati in Agricoltura nelle Regioni del Sud Italia*. Medici Senza Frontiere.

Muench, S., Stoermer, E., Jensen, K., Asikainen, T., Salvi, M., & Scapolo F. (2022). *Towards a green and digital future*. European Union. https://data.europa.eu/doi/10.2760/977331.

Muñoz de Bustillo Llorente, R. (2024). *A critical review of the digital and green twin transitions: Implications, synergies and trade-offs*. European Commission. https://publications.jrc.ec.europa.eu/repository/handle/JRC140036

OP e ANICAV. (2020). Disciplinare Di Produzione Integrata Pomodoro da Industria Centro Sud Italia, Anno 2020. http://www.apofoggia.it/public/articoli/documenti/DPI_pomodoro_da_industria_2020.pdf

Palumbo, L. (2024). *Taking vulnerabilities to labour exploitation seriously: A critical analysis of legal and policy approaches and instruments in Europe*. Springer Nature.

Perrotta, D. (2014). Vecchi e Nuovi Mediatori. Storia, Geografia ed Etnografia del Caporalato in Agricoltura. *Meridiana, 79*, 193–220.

Perrotta, D. (2016). Processing tomatoes in the era of the retailing revolution: Mechanization and migrant labour in Northern and Southern Italy. In A. Corrado, C. De Castro, & D. Perrotta (Eds.), *Migration and agriculture: Mobility and change in the Mediterranean area* (pp. 58–76). Routledge.

Perrotta, D., & Sacchetto, D. (2014). Migrant farmworkers in Southern Italy: Ghettoes, Caporalato and collective action. *Workers of the World, 1*(5), 75–98.

Tagliacozzo, S., Pisacane, L., & Kilkey, M. (2021). The interplay between structural and systemic vulnerability during the COVID-19 pandemic: Migrant agricultural workers in informal settlements in Southern Italy. *Journal of Ethnic and Migration Studies, 47*(9), 1903–1921. https://doi.org/10.1080/1369183X.2020.1857230

Veltmeyer, H., & Ezquerro-Cañete, A. A. (2023). Agro-extractivism. *The Journal of Peasant Studies, 50*(5), 1673–1686. https://doi.org/10.1080/03066150.2023.2218802

Ye, J., Van Der Ploeg, J. D., Schneider, S., & Shanin, T. (2019). The incursions of extractivism: Moving from dispersed places to global capitalism. *Journal of Peasant Studies*, 1–29. https://doi.org/10.1080/03066150.2018.1559834

Agricultural Crisis and Food Insecurity in Greece. Challenges in Strengthening Local Agri-food Systems

Michael Petrou

Introduction

In recent years, the convergence of multiple crises—including the economic downturn of 2008/2009, the COVID-19 pandemic, worsening climate change, and escalating geopolitical tensions—has profoundly reshaped the landscape of food accessibility and insecurity across the world. These overlapping challenges have aggravated social inequalities, pushing more people into extreme poverty and malnutrition. This highlights the urgent need to transform agricultural production systems and promote fairer food distribution locally and globally.

The COVID-19 pandemic and the war in Ukraine have had unforeseen consequences, severely disrupting global food markets. This compounded crisis has triggered surging inflation and led to food and energy

M. Petrou (✉)
Institute for Social Research, National Centre for Social Research (EKKE), Athens, Greece
e-mail: mpetrou@ekke.gr

S. Narotzky et al. (eds.), *Agricultural Extractivism in the Mediterranean Region*,
https://doi.org/10.1007/978-3-032-05599-6_10

insecurity, revealing the complex interconnectedness of places, and the overall fragility of global food security. As P. Raffard (2024) shows, consumers are now realizing that "almost unlimited access to food is not an inviolable reality" while agri-food products play a key role in state stability. Furthermore, climate change and an increase in extreme weather events are putting pressure on agricultural production systems, reducing yields and farm incomes while locking farmers into productive and economic stagnation. Compounding these concerns, the growing concentration of supply chain control in the hands of a few large agro-industrial corporations, alongside financialization, is intensifying speculative practices in food markets. These global developments are harming both local producers and consumers, further destabilizing the already fragile regional and local food supply chains (Woods, 2011).

In this challenging context, international organizations for agricultural and rural development, along with policy makers, after having long endorsed the industrialized, productivist model to achieve food security, are now questioning its effectiveness and sustainability. In Europe, the Common Agricultural Policy (CAP) has introduced the "Farm to Fork Strategy" within the Green Deal framework—an initiative aimed at creating a fair, healthy, and environmentally sustainable food system. This strategy emphasizes more ecological schemes of agricultural production as a significant step away from purely output-driven frameworks (European Commission, 2020). At the same time, global food movements and non-governmental organizations advocating for equitable and sustainable agriculture are intensifying their struggles against neoliberal market policies and the corporate interests of agribusiness (Andrée et al., 2014; Byaruhanga & Isgreen, 2023). They call for new agricultural policies that will ensure sustainable production, distribution, and consumption—offering an alternative to the dominant neoliberal food regime. Emphasis is placed on the right of local communities to reclaim control over land and natural resources, along with equitable access to healthy food—bringing the relocalization of agricultural production systems and sustainable food futures to the forefront of the debate (Trauger, 2014; Wittman et al., 2010).

Greece, a Mediterranean country on the capitalist periphery of Europe, has largely followed the trajectory of post-war economic and agricultural modernization (Paoli et al., 2017). Despite the dominance of small-scale farming and fragmented landholding—with an average farm size of approx. 5 hectares—family farming has undergone significant modernization and demonstrated noteworthy competitive performance (Kasimis &

Papadopoulos, 2013; Guiomar et al., 2018). Key drivers of their integration into the capitalist agricultural model included public land improvement projects, accessible agricultural credit schemes, and the subsidy-protectionist framework of the EU's Common Agricultural Policy (CAP). From the 1990s onwards, the availability of abundant cheap labor from the Balkans and other third (non-EU) countries further supported the viability of smallholdings, enabling them to stabilize and consolidate (Kasimis, 2008). However, the 2009 financial crisis revealed long-standing structural weaknesses in Greek agriculture. The fiscal austerity measures imposed under the Troika's memoranda led to the bankruptcy and exit of many vulnerable farms (Anthopoulou et al., 2017). Since then, emerging new risks and escalating geopolitical tensions continue to threaten family farming resilience, contributing to further rural depopulation and the abandonment of agricultural activities. Disillusioned by declining prospects, many farmers are either leaving the sector altogether or allocating their land for more profitable, albeit non-productive, uses, such as solar farms, when suitable opportunities arise (Petrou, 2018).

To address the aforementioned issues, this chapter draws on extensive field research conducted in the Greek countryside. It focuses on a dynamic lowland area of intensive agriculture in Beotia Prefecture—specifically the community of Drosopigi, located approximately 100 kilometers northwest of the Athens metropolitan area. The study centers on the historically pivotal role of farming in the community's social and economic fabric. It explores the processes of agricultural modernization, Drosopigi's function as a key supplier of agri-food products to greater Athens area markets, the critical role of foreign labor since the 1990s within local agricultural systems, and the emerging challenges posed by potentially profitable non-agricultural land uses against ongoing global economic volatility.

Through the experiences of local farmers, the chapter examines (a) the manifestations of the current multidimensional crisis (economic, land grabbing, energy, professional identity) and (b) how producers cope with challenges in the production process—such as dwindling incomes, land loss, the crisis of rural identity, the abandonment of parental plans, and obstacles to succession in the agricultural profession—and how they negotiate the difficulties of agricultural production system relocalisation.

From Global Crises to Local Production Impasses in Greek Rural Areas: Advocating for the Relocation of the Agricultural System

Recent global food crises are the compounded result of structural weaknesses in the industrial agricultural model, grounded in natural resource extraction and shaped by neoliberal market globalization. This extractivist model—characterized by mechanization, heavy use of agrochemicals and irrigation, land concentration, and profit-driven monocultures—has profoundly restructured rural landscapes and farming systems at regional and local scales during the post-war period. The intensification of agriculture has integrated local farms into global agro-industrial chains, expanding corporate control across all stages of agri-food supply chains—from seeds and inputs to processing and distribution—deeply altering traditional family farming (Woods, 2011, pp. 129–144).

Across Europe—and particularly in the Mediterranean region, including Greece—farming has been progressively restructured into a capitalist enterprise. This shift has contributed to the outmigration of family members' labor and a significant decline in the primary sector's role in rural economies, employment, and livelihoods. The ongoing process of de-agriculturalization reflects not only a reduction in agricultural activity, but also a broader erosion of rural socio-economic systems (Kasimis & Papadopoulos, 2001). As agricultural production and rural values lose their foothold in local communities, family farms increasingly struggle to reproduce themselves both economically and socially. In response to economic pressure, many households have turned to income diversification strategies. Others have ceased farming altogether due to diminished productive capacity or an inability to ensure long-term sustainability (Paoli et al., 2017). The overall consolidation of land into larger, mechanized farms—driven by a market-oriented logic aimed at maximizing efficiency and surpassing traditional food security goals—has had profound social and cultural impacts. Structural changes have resulted in the dislocation of agriculture from rural communities and the disconnection of consumers from the places where food is produced, eroding the cultural and social ties linking them to rural identities.

While post-war productivist agriculture initially succeeded in ensuring food security, by the 1980s it faced mounting criticism due to structural inefficiencies and budgetary distortions, environmental degradation, and deepening socio-spatial inequalities—especially in disadvantaged areas and among smallholders. The 1992 reform of the EU's Common Agricultural Policy marked a pivotal shift, introducing decoupled subsidies, agri-environmental measures, and diversification incentives aimed at promoting sustainability and reducing dependency on intensive production models. The transition towards post-productivism, reflected in the emergence of "new ruralities" has redefined rural spaces as multifunctional and regenerated, moving beyond agriculture-centered development (Mormont, 1990; Woods, 2010). More recently, the EU's Farm to Fork Strategy seeks to build more sustainable agri-food systems by shortening food supply chains between the field and the final consumer, promoting responsible farming practices, and ensuring fair income for local producers. In this context, small and medium-sized farms adopting sustainable methods are well positioned to benefit, through direct financial support, improved market access, reduced reliance on intermediaries, and long-term gains such as healthier soils, lower input costs, and higher-quality outputs (European Commission, 2020).

In Greece, food security and agricultural modernization were initially shaped by strong state intervention and later bolstered by the country's accession to the European Economic Community in 1981, particularly through CAP subsidies. Despite remarkable agricultural intensification and integration into capitalist markets, Greek farms remain small—averaging just 5.3 hectares, compared to 17.1 in the EU-27, 26.1 in Spain, and 69.6 in France according to EUROSTAT data from 2020 (Makantasi & Valentis, 2024). This reflects the resilience of family farming, sustained by kinship networks, diversified livelihoods through farmer pluri-activity, and the protective CAP framework—until the 1992 reform, which marked a shift towards market liberalization. Declining CAP support and neoliberal deregulation coincided with the arrival of abundant migrant labor from the Balkans and Eastern Europe, which proved crucial to maintaining family-based farming systems, by lowering labor costs, often through hyper-exploitation. However, the 2009 sovereign debt crisis and ensuing austerity measures imposed new hardships: agricultural land taxation, soaring input costs and overall production expenses, a cash-flow crisis combined with restricted credit access, and weakened consumer demand. These adverse developments ushered rural areas into a period of economic stagnation, bankruptcy of indebted agricultural holdings, and production impasses. Similarly, vulnerable groups faced malnutrition and food insecurity, while national food sovereignty was severely undermined due to disruptions in supply chains (Konstantinidis, 2022).

Nearly a decade of austerity and policies dictated by Troika memoranda failed to address Greece's structural production problems or correct persistent market distortions, particularly in the primary sector. These measures also lacked a coherent strategy to revitalize the rural economy and strengthen local agricultural systems. The growing imbalances in the primary sector, as evidenced by soaring food prices and conjectured shortages, are symptomatic of a globalized agro-industrial model.

Greek farmers were already facing a fragile situation when new external threats emerged in the form of the COVID-19 pandemic and the war in Ukraine. Though distinct in origin, both severely disrupted global and local food supply chains, driving speculative increases in input costs, energy prices, and retail food prices. Extreme poverty and malnutrition have spread throughout the country, including rural areas which are often considered more resilient places to live in due to food self-production and kinship and solidarity networks (Anthopoulou et al., 2025).

In this context, the strategic role of local agricultural systems for human well-being is becoming ever more apparent, positioning food access as a central social concern (Zerbe, 2014). Movements advocating for the relocalization of agricultural production systems, food sovereignty and biodiversity preservation emphasize that building sustainable local, regional, and national agri-food systems is the main challenge of our time (Burdock & Ampt, 2017; Mann, 2014). However, despite the widespread acknowledgment of the need to change the dominant model, the agricultural crisis remains deeply entrenched. Emerging challenges further complicate the transition to sustainable agriculture: worsening climate change, coupled with the depletion of natural resources—particularly water and soil—has significantly reduced yields, raising ethical concerns about environmental justice and sustainability. At the same time, the ongoing energy crisis has caused production costs to soar, making food even less affordable, intensifying issues of food insecurity and eroding state food sovereignty. These pressures exacerbate the productive and economic constraints that producers face, thereby hindering progress towards meaningful reform.

In addition, the issue of climate change—particularly its impact on agriculture—has introduced new actors and competing claims on agricultural land within the context of a "low carbon rural future" (Corson et al., 2013). Renewable energy sources, especially wind and solar, widely promoted as forms of green development, are in fact undermining farming sustainability, leading to land fragmentation and conflicts within local communities. Farmland is increasingly treated as an investment opportunity for capital accumulation, as multinational companies and non-agricultural actors gain control over land and natural resources through market environmentalism (Harvey, 2006; Bakker, 2005). This process, encapsulated as "green grabbing", is embedded in global capitalist interests that legitimize the dispossession of local farmers (Fairhead et al., 2012). By instrumentalizing the environment, these developments displace rural populations and reduce land available for food production, bringing food security and food sovereignty back into academic and political debate.

In Greece, the abandonment of agricultural land by smallholders, the bankruptcy of indebted farms, and the outmigration of younger rural residents all contribute to a landscape of rural decline and socio-economic inertia. As farmland becomes increasingly unviable for sustaining livelihoods, it is rendered susceptible to conversion to non-agricultural uses aligned with extractivist logics of social reproduction. These dynamics reflect the broader implications of intersecting crises on agricultural production, land-use transformations, and the emerging challenges of relocalizing agricultural production systems —issues that are further examined in the following section.

Productive Dynamics, Socio-economic Transformations, and Challenges of Local Family Farming. The Community of Drosopigi

The field research was conducted in Drosopigi, a fertile agricultural community in Beotia region.[1] I have been familiar with this area for many years, as it was the field site for my doctoral research (2000–2005). During that period, I spent 18 months residing in the community. The relationships established with the locals have been maintained to this day, as I have regularly returned to document ongoing transformations and update my research. The material presented in this chapter was gathered during three visits between November 30, 2024 and January 12, 2025. Each visit spanned several days and involved participatory observation and 15 open-ended interviews with agricultural producers operating small, medium, and large-sized farms with varying levels of modernization.

[1] The name "Drosopigi" is a pseudonym. It is used to preserve the anonymity and privacy of the people with whom I lived for almost a year and a half and who I still visit today.

According to the 2021 Population Census (Hellenic Statistical Authority, 2021) Drosopigi has 461 permanent residents and average landholdings around the national mean (5.3 ha). Agriculture remains the main economic activity and a cornerstone of social life in Drosopigi. Located in a lowland region, the community exemplifies an intensive model of agricultural production shaped by post-war modernization policies. Furthermore, a deep engagement with the land has endured since the community's establishment in the nineteenth century, shaping subsistence patterns, local identities, and kinship structures. Farmland for local community is more than a productive asset; it constitutes a vital social and symbolic resource. The family plot carries a particular form of sacrality, embedded in both personal and collective memory as a site of intergenerational labor, struggle, and achievement. Through these landscapes family histories are lived, recalled, and reproduced. This intimate connection between land and lineage is further reflected in the area's informal toponymy: distinct zones within the agricultural plain bear the names of local families. These vernacular geographies serve as mnemonic anchors, encoding genealogies of labor and territorial belonging into the materiality of the land. As such, the agricultural landscape functions not merely as a site of production, but also as a spatialized repository of the community's social history—where memory, kinship, and place intersect in the ongoing reproduction of local life (Petrou, 2012; Gefou-Madianou, 1992).

The integration of the area's production systems into the broader circuits of a supra-local agricultural market was facilitated by its spatial proximity to the greater Athens area. This geographic advantage—combined with the increasing demand for fresh vegetable products—enabled local producers to begin specializing in the cultivation of crops such as carrots, tomatoes, onions, and potatoes since the mid-1970s. This shift marked a significant transition from extensive, low-yield cereal cultivation to intensive, market-oriented production systems. Over time, Beotia's plain gained prominence for its horticultural crops, becoming a key supplier to Athens's retail and wholesale markets. This productive transformation was followed by notable social restructuring within the community. In particular, the shift to horticulture was initially driven by small-scale farmers (2–5 ha) who were willing to take risks that larger landholders often avoided. For them, it was a choice based on a family survival strategy—that is family labor reserve, reinvestment of savings in mechanical farm equipment, drilling, and land purchase or rent—in a demanding competitive context (Petrou, 2012). As one producer, Vasilis, now 76, recalls:

> We, the hungry, dared. When you're not hungry, you don't take risks. The big farmers who had 500 acres [50 ha] of wheat and barley had a certain income, so there was no reason for them to try something new. (December 2, 2024)

The high demand for vegetables generated significant profits for producers, prompting many of them to invest in markers of social mobility and status—such as land, modern machinery, cars, and urban-style housing. The successful engagement with new, dynamic crops came to redefine local hierarchies not only in terms of economic capacity, but also regarding social recognition. Small-scale farmers, once perceived as marginal in the local hierarchy, were now increasingly viewed as "worthy and successful producers," challenging long-standing assumptions that only large landowners could achieve material prosperity and social status (Marsden, 1999; Fuller, 1990).

Another critical milestone of structural transformation that significantly reshaped both economic dynamics and socio-cultural landscape in Drosopigi was the influx of a large, devalued migrant labor in the early 1990s, originating from the Balkans (primarily Albania and Romania) and South Asia (notably Pakistan and India). The availability of abundant and unregulated labor, combined with natural resources—particularly a rich aquifer and fertile soils—significantly influenced producers' cultivation strategies. Many shifted to labor-intensive crops such as peas and green beans, which required manual harvesting. As a result, the total area under vegetable cultivation expanded markedly within the community: from 200 hectares out of 1600 in 1991, to 380 hectares in 2001, reaching 450 hectares on the eve of the country's financial collapse. This transformation was accompanied by the development of post-harvest infrastructure, such as packing plants, as well as an expansion into wholesale markets and new supra-local professional agreements of farmers with large food distribution chains and supermarkets based in the greater Athens area. These developments led to substantial profit margins. All categories of producers, whether operating small holdings (2–5 ha), medium-sized farms (5–10 ha), or larger operations (over 10 ha), benefitted from the availability of migrant labor. This evolution marked a reconfiguration of rural identity: from the traditional smallholder farmer of the early 1970s to a small entrepreneurial farmer-merchant-employer, managing a business capable of employing dozens of foreign workers.[2]

The Dynamic Micro-capitalist Farm in the Wake of the Multiple Crisis

However, this dynamic trajectory of micro-capitalist family farming was disrupted in the late 2000s with the onset of Greece's financial crisis and the imposition of fiscal restructuring and austerity measures. Farmers were confronted with a number of restrictions and production impasses. Yet, the economic turmoil is not over, as the local community has been affected by global crises such as the increase in energy and production costs resulting from the war in Ukraine and the COVID-19 pandemic. These combined crises have exacerbated insecurity and fostered distrust towards promises made by the government to address problems in the primary sector. As Petros, a 45 year old small-to-medium capacity potato producer, mentions:

> We barely got over the memoranda and covid arrived. Along with the bans on moving agricultural products between prefectures; closed restaurants, closed eateries, closed hotels. And you're expected to grow crops and sell them. Now we have the energy crisis and we don't know how to protect ourselves. (December 2, 2024)

The ongoing energy crisis and the rapid rise in production costs—fuel, agricultural inputs, electricity, and more—have pushed many farmers to the brink of collapse. Particularly affected is the once-promising and dynamic "middle class" of producers with significant productive and business achievements which was expected to respond and successfully withstand the successive crises. Nevertheless, it appears to stumble and is faced with the risk of overturning the production and consumer model they had embraced. Their long-standing practices of expanding into large-scale

[2] The presence and employment of hundreds of migrants in the community's farming obviously raises a series of interesting aspects about cohabitation relations with locals (labor relations, housing conditions, migration patterns, cultural classifications and evaluations on both sides, etc.). However, the study of these issues goes beyond the objectives of this article. More details on these issues as well as on the cultural dimension of the transformation of the farmers' identity, namely from simple cultivator to boss/employer, see M. Petrou (2008, 2012).

farming, which are not always based on sound economic rationale, but rather on short-term profit opportunities, the pressure to respond to local competition, as well as a tendency towards ostentatious consumption to enhance social status (e.g. expensive cars, trips abroad), are now proving unsustainable. These patterns, once seen as symbols of success and sources of recognition within local communities, are increasingly undermining the very identity of the "entrepreneur-farmer" or "local boss." The cultural and economic standing they had built is now at risk, revealing deeper vulnerabilities in both the agricultural economy and the rural social fabric.

Nikos, 60 years old, falls in this category of producers who had a successful business career but are currently experiencing significant production and financial setbacks. He points out:

> Successive crises have brought us to our knees. Costs have skyrocketed. From the 20 hectares of vegetables I used to grow, I now plant 3 hectares. We thought this happiness would last forever and we didn't think about tomorrow. I'm barely getting by. (December 3, 2024)

Regardless of the poor choices made by some farmers, the multifaceted agricultural crisis is also affecting producers who have maintained a more careful and rational management of their farms. As Giannis, 59 years old, one of the most dynamic producers in the region, mentioned during a discussion at the village café:

> Last year, basic fertiliser cost me 22 euros and this year it cost me 52. Last year, for the whole year, I spent 20 thousand euros on petrol and this year, for the same land, I've spent 40 thousand euros on petrol. (December 4, 2024)

At that point, he took his mobile phone out of his pocket and showed me a message he had received a few days prior with his electricity bill for his borehole:

> There, look, 11,000 euros for a month. Just for a month. Last year, for the same borehole, over the same period, I paid 3000 euros. Out of the 100 boreholes that operated in the entire village last year, this year only 40 are operating because of the cost. But I can't not water my vegetables. They need a lot of water because of the drought of recent years. I'm using all the savings I had on the side to pay for electricity, petrol, fertiliser. But how long can I last with these costs? (December 4, 2024)

While my interlocutor was explaining all the above, he was interrupted by the presence of a man with a briefcase approaching café customers, looking for a local farmer's house. "Bailiff, bailiff" my interlocutor whispered, adding: "*mortgaged fields, mortgaged boreholes, mortgaged warehouses, people have reached their limits.*"

The economic and energy crisis shook not only medium and large producers, but also the majority of small ones, such as Sofia, a 45 year old wife of a livestock farmer. They were forced to slaughter and sell several animals from their farm for meat as it was deemed much more profitable than raising the livestock. Sofia highlights an additional dimension of the crises, that of fiscal-tax policy:

> With the prices of animal feed soaring, we can't afford to have a herd. We would be able to cope though if there were a generous cut in the tax on diesel. It costs 0.75 euro and with taxes it comes to 1.40 euro. Meaning that, almost half the price is tax. It's just impossible. If they cut the tax on petrol, we'll be able to grow our own animal feed, barley, lupines, peas, broad beans, which thrive here, and we won't need traders and imported animal feed. (December 10, 2024)

The development of local crops mentioned by Sophia could offer a solution to the production and financial impasses faced by producers in the community. The global disarray in distribution channels and the shortages in markets of nutritionally important foods, such as wheat, due to the war in Ukraine could be mitigated by promoting the domestic production of wheat, as part of a broader strategy of relocalizing agricultural systems to shorten provisioning chains. . Wheat is a crop that has long thrived in the area and its cultivation would seem to confer significant benefits to producers due to the geopolitical situation. However, wheat cultivation in the community has declined rapidly (as well as across Greece) because of it sunprofitability. This decline is driven by several factors, including the decoupling of CAP subsidies and competition from cheap imports from third countries or EU nations with lower production costs, such as Ukraine and France. Although the war in Ukraine has increased global demand for cereals, producers are not benefitting from higher prices. This is largely due to pressure exerted by wholesalers on the producer price coupled with increased speculation in the domestic market, undermining any potential gains for local producers. George, 61 years old, a wheat producer, explained:

> It's no longer profitable to grow wheat. The basic cost two years ago was 550 euros per hectare and today it is 1500 euros. The yield per hectare is 3000 kilos and last year agri-businesses bought it for 0.50 euro per kilo. If you work it out, I only covered my expenses. I don't include my toil my anxieties and my wages. This year I sold it to an agro-industrial company and 4 months later not only have I not been paid, they haven't even given me a price yet. That's why farmers are quitting. There are 1600 hectares in the village and this year only 700, fewer than half, were cultivated. If this situation continues for another 2–3 years, then there will be suicides. I, for one, will quit. (December 11, 2024)

In this context, the successive recent crises have aggravated the difficulties in adapting and restructuring, not only for small farmers who are more vulnerable to price variations and foreign competition from cheaper imported vegetables (mainly from Egypt and Turkey) but also for larger farmers who indebted themselves in order to invest in technological modernization, innovation, and marketing of their agricultural holdings. There are numerous cases of producers who took out loans to invest in new crops, and for the purpose of family succession, such as Sotiris, 65 years old, one of the most dynamic winegrowers in the community. He decided to stop growing tomatoes and wheat and focus exclusively on viticulture in order for his son to manage the holding. However, this decision is being called into question, not only by himself but by his son as well:

> The year before last, the winery bought grapes off us for 30 cents, last year for 28 and this year, 26 cents per kilo, while the cost of producing has tripled. Every year it gets worse. We're not doing well at all and I have my son's doubts, nagging and arguments, and he, rightly, says: 'Things aren't looking good. So much investment wasted and no profits? You presented and calculated things differently when I started out.' (December 17, 2024)

The cases of the two producers mentioned above illustrate how large agri-food companies profit from fluctuations in prices they set themselves. In the case of the region's flour mills and wine industries, the downward trend in the farm gate producer price of wheat and grapes is not accompanied by a corresponding drop in the price per kilo of bread and wine on the consumer market.

The current crisis in the region is unfolding in an even more palpable way in the recent labor shortage as the daily wage for foreign workers has risen from 25 euros to 35 or 40 euros—a cost that many producers cannot afford—pushing many workers to move to other areas in search of work, thus deepening the labor shortfall locally. This shortage has caused discord and conflicts among producers as some are willing to pay a little over the local average wage to attract workers, leading to strong reactions, disputes, and even disruptions to long-term friendships and relationships among producers.

It is important to note, however, that this labor shortage has affected the entire Greek countryside in recent years. The underlying causes lie in the poor living and working conditions faced by foreign agricultural laborers, including low wages, insecure employment, inadequate housing, and the deterioration of labor rights. These issues are compounded by the state's persistent failure to implement effective policies that support the professional and social integration of foreign workers. Furthermore, there has been little effort to ensure the safe and organized movement of seasonal migrant laborers across regions, leaving both workers and producers vulnerable (Kapsalis, 2022).

Another limiting factor to the management of crops is the disruption in the bureaucratic procedures of the public services implementing agricultural policy. The dismissal of thousands of public servants as dictated by the memoranda imposed on Greece has thrown many public services into disarray. Katerina, 45 years old, a farmer who recently lost her husband and a mother to three children, specializes in the cultivation of industrial tomatoes (pomodoro) and like many of her colleagues is faced with a major challenge. She experiences melancholy, emotional instability, sometimes confusion, and a generalized apathy and personal anguish that verge on depression. She often falters and finds it unbearable to live in an environment with no stability, no prospects, and rapid reversals of the certainties that she had relied on until the onset of the crises era in the late 2000s:

> In the summer I planted 4 hectares of industrial tomatoes. In June hail destroyed it all. There was nothing left. I lost 40,000 euros, all my toil. Five months later the authorities still haven't come to certify the damage for me to be compensated. In the meantime, I ploughed the fields in preparation for next year. I kept them with the destroyed tomatoes for as long as I could for them to see the damage, but I couldn't wait any longer. I have called them so many times, for them to come, but there are only two employees for the whole region and not enough time. So when they eventually come, what will they be able to certify? Will I be compensated? Will I be paid? In the meantime I owe money to agronomists, to the power company, to

suppliers. What am I going to do? How am I going to manage? (December 18, 2024)

In light of the above, it can be argued that the evolving landscape of national and supranational political and economic developments is ushering in a new era of business restructuring and growing productive imbalances. Many producers are confronted with a new reality marked by financial constraints and heightened uncertainty regarding their production goals and income stability. Producers are now being compelled to abandon long-standing and deeply embedded productive mentalities and practices—historically associated with extensive land cultivation, traditional entrepreneurial activity, and a firmly established professional identity centered on the figure of the "boss"/entrepreneur. The multidimensional crisis of recent years has led to the questioning of the dominant production model and the identity of the farming profession for the first time. The energy crisis, in particular—with soaring costs for fuel, agricultural inputs, electricity, and more—has placed enormous strain on production viability. As a result, many farmers are now struggling to survive. For many, adapting to this new reality proves difficult. Success, recognition, and, above all, the social prestige associated with the long-held "boss" status are increasingly under threat. While their current productive capacity may still afford them a basic livelihood, it is no longer sufficient to sustain the once-dominant production and consumption model they embodied—one through which they derived power and social distinction in the context of a micro-capitalist enterprise.

Towards a New Agricultural Extractivist System. The End of Agriculture in Drosopigi?

At the same time, the economic crisis of 2009, which created severe operating problems and production impasses for farms in the area, has facilitated the move to non-productive uses of land. An increasing number of farmers are selling or leasing their lands for solar panel installation, leading to a "service agriculture." The dominant agricultural production identity has been shaken by the installation of solar farms, currently occupying an area of 200 hectares of fertile, productive land, provoking reactions and conflicts within the local community (Petrou, 2018). The effects can be particularly detrimental and unavoidable, especially where the current multidimensional crisis has weakened local resistance to the proposals and pressures of non-local, mostly urban, actors regarding the siting and development of non-agricultural infrastructure that financially benefits some members of the local community.

Active farmers, along with several former internal migrants who no longer have a stable relationship with the community, are involved in this "game" of selling agricultural land. These individuals—either former residents or their descendants—had previously migrated from the area for education, work, or marriage, often relocating to major cities, especially Athens. Devoid of strong emotional ties to their ancestral land, these migrants more easily cede their agricultural land for alternative uses that provide them with a steady income. They perceive the shift in land uses—especially in the context of ongoing crisis—as a favorable opportunity to make a profit given that the price of land designated for industrial use is much higher than for agricultural use. In this case, land is regarded as a means for opportunistic enrichment. Notably, it emerges that the actors contributing to the fragmentation of the agricultural landscape and the degradation of the natural environment are not necessarily "outsiders" such as photovoltaic entrepreneurs, but may also be locals.

A typical case is that of 55-year-old Nikos who grew potatoes, onions, beans, and wheat for years on 15 hectares of land. This year, for the first time in the history of the family farm which he inherited from his father, he decided to significantly alter their production model. With a sense of relief and reflection, he explains:

> This year I haven't planted any beans, or potatoes, or wheat. Cultivating them over the last 5 years has been unprofitable. And even if I did well one year, the next year I would make a loss. So overall, I would break even. The only thing I covered were my expenses. And I don't include the torment, the running around, the fatigue. I stopped and I've found peace. I'm done. I've only kept 5 hectares for onions because I made a deal with a trader and I can't back out, and because they sell for a good price. The other 10 hectares I've leased to a solar panel company for 12,000 euros per year. Meaning that I have 1000 euros in my pocket, guaranteed every month, which I couldn't make with potatoes, beans and wheat. But, I wonder, with fields worth 300,000 euros, tractors, tools, machines and warehouses, all worth that much again, plus the knowledge I have gained. If I can't cope, then there is something wrong with farming. (January 9, 2025)

In the context of ongoing and compounding crises, coping with current harsh conditions has become increasingly difficult for many farmers.

As a result, a growing number are choosing to reduce their agricultural activities and seek alternative sources of income. This weakening of agricultural production, combined with the promising financial prospects offered by non-agricultural land uses, has become a lure for many producers facing economic and structural challenges, encouraging them to exit the agricultural sector. The phenomenon of part-time employment and the pursuit of supplementary non-agricultural income are not new in the rural world. The pluri-activity of farmers—whether as freelancers in construction, tourism, or trade, or as occasional employees—has long been a structural component of the resilience of Greek agriculture. Despite the small size of the family farms, the sector has managed to modernize and withstand competition in international markets, also due to the strong subsidy support of the CAP (Kasimis, 2008; Kasimis & Papadopoulos, 2013). What is novel in the case of the community of Drosopigi, however, is the emergence of renewable energy development as a catalyst for a radical and permanent shift in land use.

Productive agricultural land—previously dedicated to food production—is now being converted to industrial use in a rural area that plays a key role in supplying the metropolitan consumer market. This transition undermines national food security and introduces additional environmental risks, particularly in the context of climate change and the reduction of cultivated land. Beyond these immediate impacts, the shift also signals a deeper transformation: an advanced de-agriculturalization of the countryside, which is losing its fundamental productive resource. This is accompanied by significant changes in rural landscapes and in the functions of rural areas. Once primarily spaces of production, these areas are increasingly becoming zones for service provision, in line with the broader trend of rural tertiarization (Woods, 2011).

Undoubtedly, the regional economy is becoming increasingly less reliant on agriculture and increasingly dependent on non-local economic actors who invest substantial capital in industrial infrastructures that are disconnected from the area's traditional agricultural functions. This shift is not inherently detrimental to local development—particularly when it arises through participatory planning processes, broad social consultation, and the small-scale investment of local energy communities. In fact, such investments were, to some extent, anticipated, given that projects of this scale require expansive land areas that are often unavailable in urban environments. However, what is concerning is the siting of these developments on land with high agricultural productivity—land that remains unprotected by the national institutional framework, in the wake of memorandum obligations that imposed fast-track procedures regarding the environmental impacts of large projects in order to attract foreign investment for the country's economic recovery. The preservation of such land is vital for ensuring the country's food sovereignty. Beyond the economic and institutional implications, these non-agricultural investments risk causing irreversible disruptions to local ecosystems and diminishing the aesthetic and ecological value of rural and natural landscapes. This includes areas of environmental and cultural significance, such as mountains, forests, and traditional agricultural territories (Ingold, 2000, pp. 189–218).

The most important concern, however, is their structural transformation into mono-themed parks, which risks leading to the "deterritorialization of cultures" and the emergence of "non-places"—spaces lacking identity, meaningful relationships, and historical connection for their users (Augé, 1995, pp. 129–175). In Drosopigi, such developments threaten to erode the deep-rooted ties that residents have with their land, their community, and their history—bonds forged through generations of family labor and the collective construction of both material livelihoods and personal identities since the 1970s, during the agricultural modernization years.

Therefore, in the context of the economic status quo and capitalistic expansion, the thoughts of a local, Stefanos, who is 80 years old today and neither supports nor completely opposes new land uses in the area, are perhaps indicative of local survival dilemmas:

> We need to realize that farming is dying out. We need to really consider what we want for our villages. Do we want our areas to remain untarnished and unpolluted by industry and outsiders? or do we think of the financial impasse we are in and the people who have needs and are selling their land in order to make a living. (January 10, 2025)

Conclusion

Over the past 15 years, a series of economic and geopolitical crises have disrupted global food supply chains, unveiling deepening food insecurity and the erosion of food sovereignty at both regional and national levels. These vulnerabilities are further exacerbated by the accelerating impacts of climate change, declining agricultural yields, and the steady conversion of farmland to more lucrative non-agricultural uses—as local farmers seek ways to survive or to simply have more profitable livelihoods. These trends are reinforced by neoliberal development paradigms that prioritize market-driven solutions, often concealed under the dominant green development discourse such as renewable energy projects. This example reflects another dimension of land-resource exploitation and extractivist economy; the green grabbing phenomenon, which is taking on worrying proportions within the EU (Cotula, 2014). While the European CAP's Farm-to-Fork strategy aims to the transition towards more sustainable food supply chains by strengthening agricultural holdings, the European Green Deal promotes green investments that can, paradoxically, undermine sustainable use of land intended for food production. In this context, rethinking the relocalization of agricultural production systems emerges as a critical and urgent challenge.

Through the ethnographic example of Beotia, the experiences of local farmers reveal a complex interplay of structural market distortions, inconsistent rural development policies, and institutional shortcomings in the protection of agricultural land. For many farmers, high production costs and fuel surcharges have rendered unprofitable crops that had thrived in the region for decades and constituted the main source of income for rural households. This has led to trends of abandoning farmland or maintaining it with a wait-and-see approach among farmers, pending future developments. It has not only compromised production capacity but also discouraged future investment, generating a pervasive sense of uncertainty and insecurity. Even farmers who have decided to invest in promising crops, such as peas and vines, faced significant risks due to the lack of a regulatory and price control mechanism that would protect them from the speculative practices of dominant agri-food monopolies. The resulting erosion of agricultural identities further weakens the cultural and material foundations of rural life. Such impasses also undermine farmers' ability to resist land-use changes towards more profitable non-agricultural uses, limiting available arable land and, consequently, the potential adequacy and strengthening of agri-food systems. The pursuit of national food sovereignty and the relocalization of agricultural production systems by maintaining and strengthening local farms is a recurrent challenge across all contexts.

The agricultural landscape in the area is increasingly losing coherence as a production system perspective, as many producers no longer regard farming as their primary professional activity or main income source worth investing in. Remarkably, this development is not happening in a remote, resource poor, and depopulated area with outdated agricultural production systems, but in a fertile plain, endowed with abundant water resources, modern mechanized farming, and a skilled labor force—located in close proximity to a major urban center. The wider Beotia area plays a considerable role in supplying fresh produce to the Athens metropolitan area and contributes significantly to national food sovereignty. However, despite the adequacy of social, economic, and natural capital, the area has become extremely vulnerable and dependent on global dynamics, undermining the viability of locally driven development.

This chapter advocates the relocalization of agricultural production systems with a view to strengthening food sovereignty at regional and national levels. Importantly, this perspective does not imply a return to small-scale, traditional or non-industrial agriculture, primarily using agroecological practices. It rather refers to the capacity of local micro-capitalist family farms (Guiomar et al., 2018) to define their ownproduction system based on good agricultural practices (Gordillo & Mendez Jeronimo, 2013) and assert control over land, production, markets, and ecological resources. In the case of Drosopigi community and Beotia plain, maintaining family farms plays a vital role in preserving the region's productive capacity, social fabric, rural landscapes, and cultural identity. These farms also contribute to reinforcing urban–rural linkages and revitalizing rural society and the local economy.

References

Andrée, P., Ayres, J., Bosia, J. M., & Massicotte, M.J. (2014). Introduction: Crisis and contention in the new politics of food. In P. Andrée, J. Ayres, M. J. Bosia, & M.J. Massicotte (Eds.), Globalization and food sovereignty: Global and local change in the new politics of food (pp. 3–20). : University of Toronto Press.

Anthopoulou, T., Kaberis, N., & Petrou, M. (2017). Aspects and experiences of crisis in rural Greece: Narratives of rural resilience. *Journal of Rural Studies, 52*, 1–11. https://doi.org/10.1016/j.jrurstud.2017.03.006

Anthopoulou, T., Partalidou, M., Nikolaidou, S., & Stamatopoulou, G. (2025). Poverty and housing precariousness in rural areas under the lens of social resilience. Revealing societal implications of lived experiences of rural poverty. *Societal Impacts, 6*, 1–14. https://doi.org/10.1016/j.socimp.2025.100119

Augé, M. (1995). *Non-places: Introduction to an anthropology of supermodernity* (J. Howe, Trans. from french). Verso.

Bakker, K. (2005). Neoliberalizing nature? Market environmentalism in water supply in England and Wales. *Annals of the Association of American Geographer, 95*(3), 542–565.

Burdock, R., & Ampt, P. (2017). Food sovereignty: The case and the space for community led agricultural autonomy within the global strategic sramework for food security and nutrition. *Journal of Agricultural Science, 9*(5), 1–18.

Byaruhanga, R., & Isgreen, E. (2023). Rethinking the alternatives: Food sovereignty as a prerequisite for sustainable food security. *Food Ethics, 8*(2), 1–20.

Corson, C., Mac Donald, I. K., & Neimark, B. (2013). Grabbing 'green': Markets, environmental governance and the materialization of natural capital. *Hyman Geography, 6*(1), 1–15.

Cotula, L. (2014). Addressing the human rights impacts of "land grabbing". D-G for external policies of the Union. EXPO/B/DROI/2014/06.

European Commission. (2020). Communication from the Commission to the European Parliament, the Council, the European Economic and Social Committee and the Committee of the Regions: A Farm to Fork Strategy for a fair, healthy and environmentally-friendly food system COM/2020/381 final.

Fairhead, J., Leach, M., & Scoones, I. (2012). Green grabbing: A new appropriation of nature? *The Journal of Peasant Studies, 39*(2), 237–261.

Fuller, A. (1990). From part time farming to pluriactivity: A decade of change in rural Europe. *Journal of Rural Studies, 6*, 361–371.

Gefou-Madianou, D. (1992). *Alcohol, gender, and culture*. Routledge.

Gordillo G., & Mendez Jeronimo O. (2013). *Food security and sovereignty*. FAO (Base Document for Discussion). https://openknowledge.fao.org/server/api/core/bitstreams/61f3f359-669c-40ff-88c6-b39b48135f3a/content.

Guiomar, N., Godinho, S., Pinto-Correia, T., Almeida, M., Bartolini, F., Bezák, P., et al. (2018). Typology and distribution of small farms in Europe: Towards a better picture. *Land Use Policy, 75*, 784–798. https://doi.org/10.1016/j.landusepol.2018.04.012

Harvey, D. (2006). *Spaces of global capitalism: A theory of uneven geographical development*. Verso.

Hellenic Statistical Authority. (2021). Population and housing census 2021. https://www.statistics.gr/el/statistics/-/publication/SAM03/-

Ingold, T. (2000). *The perception of the environment. Essays in livelihood, dwelling and skill*. Routledge.

Kapsalis, A. (2022). *Mobility and employment of refugee populations in the Greek region*. IOM-UN. https://greece.iom.int/sites/g/files/tmzbdl1086/files/mobility-and-employment_en_online.pdf.

Kasimis, C. (2008). Survival and expansion: Migrants in Greek rural regions. *Population Space and Place, 14*(6), 511–524.

Kasimis, C., & Papadopoulos, A. G. (2001). The de-agriculturalisation of the Greek countryside: The changing characteristics of an ongoing socio-economic transformation. In L. Granberg & I. Kovách (Eds.), *Europe's green ring* (pp.197–218). Ashgate. https://doi.org/10.4324/9781315255811-8.

Kasimis, C., & Papadopoulos, A. G. (2013). Rural transformation and family farming in contemporary Greece. In A. Moragues Faus, D. Ortiz-Miranda & E. Arnalte Alegre (Eds.), *Agriculture in Mediterranean Europe: Between old and new paradigms* (pp.263–293). Emerald.

Konstantinidis, C. (2022). Food insecurity, austerity, and household food production in Greece, 2009–2014. *Review of Social Economy.* https://doi.org/10.1080/00346764.2021.2025414.

Makantasi, F., & Valentis, E. (2024). The agricultural sector in Greece. DIANEOSIS Organization for Research and Analysis (in Greek) https://www.dianeosis.org/2024/09/o-agrotikos-tomeas-stin-ellada/

Mann, A. (2014). *Global activism in food politics: Power shift.* Springer.

Marsden, T. (1999). Rural futures: The consumption countryside and its regulation. *Sociologia Ruralis, 39*(4), 501–520.

Mormont, M. (1990). Who is rural? Or, how to be rural: Towards a sociology of the rural. In T. Marsden, P. Low, & S. Whatmore (Eds.), *Rural restructuring. Global processes and their responses* (pp. 21–44). David Fulton Publishers.

Paoli, J.-C., Anthopoulou, T., Ben Saad, A., Bergeret, P., Elloumi, M., Napoléone, C., & Vianey, G. (eds.) (2017). La petite exploitation agricole méditerranéenne, une réponse en temps de crise. Options Méditerranéennes, *Série A, 117.* Centre International de Hautes Études Agronomiques Méditerranéennes (CIHEAM-IAMM). https://om.ciheam.org/option.php?IDOM=117

Petrou, M. (2008). Immigrants, farm labor and producers. Cultural peculiarities and the journey to symbiotic relations in an agricultural community. In H. Kasimis, L. Louloudis, & E. Zakopoulou (Eds.), *Agricultural society and popular culture* (pp. 137–161). Plethron (in Greek).

Petrou, M. (2012). Rural immigration, family farm modernisation and reactivation of women's traditional farming tasks in Greece: Masculinities and femininities reconsidered. *South European Society and Politics, 17*(4), 553–571. https://doi.org/10.1080/13608746.2012.654617

Petrou, M. (2018). Diverging sociocultural visions of the rural and land use conflicts in times of crisis. Solar farm development and farmland dispossession. *Option Mediteraneennes, 117*, 167–180. http://om.ciheam.org/om/pdf/a117/a117.pdf

Raffard, P. (2024). *Géopolitique de l'alimentation et de la gastronomie. De la fourche à la food tech.* Le Cavalier Bleu Editions.

Trauger, A. (2014). Toward a political geography of food sovereignty: Transforming territory, exchange and power in the liberal sovereign state. *The Journal of Peasant Studies, 41*, 1131–1152.

Wittman, H., Desmaris, A., & Wiebe, N. (2010). *Food sovereignty: Reconnecting food, nature and community.* Fernwood Publishing.

Woods, M. (2010). Performing rurality and practising rural geography. *Progress in Human Geography, 34*(6), 823–839. https://doi.org/10.1177/0309132509357356

Woods, M. (2011). Rural geography. In *Processes, responses and experiences in rural restructuring*. Sage Publications.

Zerbe, N. (2014). Exploring the limits of fair trade: The local food movement in the context of late capitalism. In P. Andrée, J. Ayres, M. J. Bosia, & M. J. Massicotte (Eds.), *Globalization and food sovereignty: Global and local change in the new politics of food* (pp. 84–110). University of Toronto Press.

Navigating Opposition and Recognition: Changing Relationships and Imaginaries Between Agroecology and the State

Patricia Homs Ramírez de la Piscina
and Paula Escribano Castaño

Food Systems in a Crisis Context

Crises are a recurrent, structural feature of capitalism and are necessary for the capitalist system to function and expand (Harvey, 2011). In Spain, the state has played a key role in managing the successive crises in recent decades, including the 2008 mortgage crisis and the COVID-19 crisis, as

Both authors participated in the conceptualisation, analysis, drafting and revision of the chapter

P. Homs Ramírez de la Piscina (✉)
Universitat de Barcelona and Food Action Research Observatory, Barcelona, Spain
e-mail: pati.homs@ub.edu

P. Escribano Castaño
Universidad Complutense de Madrid, Madrid, Spain

S. Narotzky et al. (eds.), *Agricultural Extractivism in the Mediterranean Region*,
https://doi.org/10.1007/978-3-032-05599-6_11

it has the ability to redistribute economic resources and the authority to modify the legislative framework, including in regard to food systems.

During crises, the state has sought to restore the conditions that allow capital accumulation to continue, bolstering economic systems that benefit the elite rather than ensuring the reproduction of life and the wellbeing of the population as a whole (Harvey, 2011; Graeber, 2011). In agrifood systems, support for capital accumulation is synonymous with support for agroindustry over small-scale food systems. This is apparent when we examine subsidy schemes, which propose "budgetary imbalances imposed by sectoral productivist inertias in favour of more intensive systems" (Oñate et al., 2023, p. 4). Despite the European Common Agricultural Policy's (CAP) intention to support extensive forms of agriculture, in practice, intensive agroindustry continues to monopolise state subsidies, arable land and resources such as water (Idescat, 2024). An example of this can be found in the COVID-19 crisis in Spain, when the state facilitated the continuation of food production activities linked to agroindustry and hampered the activities of small-scale food systems. On the one hand, the restrictions on mobility and the hampering of productive activities under the "state of exception" legislation affected small-scale producers in vegetable gardens, food coops and farmer's markets, limiting their social reproduction. On the other hand, the same legislation promoted and protected the continued activity of large companies in the agrifood sector (Gascón, 2020; Batalla et al., 2020). Throughout history, state political and economic forces have shaped the capitalist development of the food sector in Spain and beyond, crisis after crisis.

Since the Green Revolution, the agroindustrial system worldwide has become far stronger than traditional small-scale systems. For example, the ten largest agrifood corporations control over 75% of the global market for commercial seeds and agrochemicals (Shiva, 2016; Ploeg, 2008). This has a major impact on the autonomy of small farms, limiting and even preventing their social reproduction on the margins of the market economy. Moreover, the expansion of globalised food systems has made farming more precarious, reducing the amount of land owned by small-scale farmers and plunging farmers into debt and economic dependence, making their farms unviable (Ploeg, 2008). The autonomous community of Catalonia in Spain, where our study is based, has seen a drastic decline in the number of farmers making a living from small-scale production in rural and periurban environments and small farms over the past 60 years, while large companies have grown and obtained more and more land. In

2018, only 1.1% of Catalonia's working population was employed in the primary sector,[1] despite nearly half of all workers in the autonomous community working in that sector at the start of the twentieth century (Pomar et al., 2018). From 2015 to 2020, the number of individuals registered as self-employed in the agricultural sector fell by 7000, while the number of companies grew (Idescat, 2024). During the same period, the net area farmed by self-employed people (linked to family farms) fell by 25,000 hectares, while the net area controlled by companies increased by 25,000 hectares.

As the industrial food system has expanded, tensions have arisen due to the negative social and environmental impacts of the Green Revolution (McMichael, 2009). Analysing the behaviour of contemporary capitalism and its crises, Nancy Fraser (2013) argues that capitalism triggers a "triple movement": the expansion of the self-regulating market, the protective reaction of society and the struggle for emancipation, with agroecology representing an example of the latter. According to Fraser, struggles for emancipation are sometimes coopted by neoliberalism, distorting them and transforming them into mechanisms that reinforce market logics instead of challenging them. Similarly, Harriet Friedmann (2005) describes the current food regime as "green capitalism" to refer to the capital restructuration process whereby environmental issues are incorporated into agroindustry. For example, there was an attempt to coopt agroecological initiatives in Madrid during the COVID-19 crisis. Agroecological producers experienced a forced "acceleration of formalisation processes," an accelerated pressure to move from the informal to the formal economy and scale up their activities. This subsumed their struggles for emancipation under the logic of capital, due to the lack of state protection for their economies and the threat of sanctions for continuing to supply food during the pandemic (Escribano, 2023, p. 4). Historically, many agroecological initiatives have operated within the paradigm of the so-called "informal economy" (e.g. consumer groups, occupation of public space for deliveries, forms of validation based on trust and mutual understanding rather than on official seals, etc.). However, whereas it was very common in the 1990s to explicitly reject state's intervention, there is now some degree of acceptance of the need for the state to guarantee agroecological sustainability.

[1] These data do not include seasonal workers or small farms that are not registered as agricultural holdings.

In this chapter, we will explore the changes that have occurred in agroecological food systems in Catalonia in relation to the state over the past 30 years. Taking an ethnographic approach, we will examine the social, economic and cultural tensions that have arisen in a context of crisis in the agroecological sector. How has the closer relationship between agroecological initiatives and the state affected alternative logics, values and socioeconomic practices? What new tensions are emerging as a result of the current positionality of agroecology in relation to the state?

Methodology

This chapter addresses two case studies. The first one analyses agroecological provisioning systems comprising small-scale farmers and food processors working alongside consumer groups, consumer cooperatives and cooperative supermarkets. We focus on the tensions and struggles to ensure their economic sustainability. In particular, we examine the difficulties that emerge from unpaid work within these networks, food prices and the role of the state. The second case study explores the tensions within agroecological initiatives in (semi-)extensive livestock farming as they attempt to resist wholesale assimilation into the market economy. It delves into the difficulties facing workers in agroecological provisioning systems when it comes to ensuring the social sustainability of such initiatives.

The data were collected through a qualitative methodology based on ethnographic fieldwork, using participant observation, semi-structured interviews and in-depth interviews (20 interviews for the livestock farming case study and 43 for the agroecological provisioning systems case study). In both cases, intermittent observation was carried out between 2013 and 2023. The interviews were recorded and subsequently transcribed. The names of the informants were anonymised in compliance with the ethics and confidentiality protocols established for the study. All data processing was carried out in accordance with rules approved by the ethics committee at the University of Barcelona.

Agroecology in Catalonia

Agroecology originated and took root in Latin America in the 1970s and 1980s as a scientific discipline, a set of peasant practices that respected all life, and a social and political movement (Sevilla-Guzmán & Woodgate, 2013). In the words of Eduardo Sevilla-Guzmán, agroecology is "the ecological management of natural resources through forms of collective social action, which offer alternatives to the current civilisational crisis" and this management of food systems is based on "participatory initiatives in the areas of production and alternative circulation of their products, aiming to establish forms of production and consumption that help address the ecological and social harms caused by contemporary neoliberalism" (2011, p. 13). Since agroecology arrived in Spain in the 1990s, transformative production and consumption initiatives have spread, such as associations of producers and consumers and consumer cooperatives (Cuéllar & Sevilla, 2013; Guzmán et al., 2000).

While there were already proposals for organised local networks in Catalonia prior to this, including several consumer cooperatives and organic agriculture initiatives, it was not until the 1990s that people began to advocate for agroecology. The expansion of this movement coincided with the introduction of government rules on third-party organic certification, that is, certification by an external agency that applies a series of technical controls to ensure that the product meets certain standards.[2] Throughout the course of this process of standardisation and institutionalisation of organic agriculture, "organic" came to be reduced to a series of production techniques, thus, losing the social, economic, political and cultural dimensions of food systems (Giraldo & Rosset, 2017). As such, third-party certification can be interpreted as a process by which organic agriculture is appropriated and simplified, as it is narrowed down to a series of regulations concerning production techniques, overlooking the sociocultural, environmental and political complexity of food systems (Homs & Narotzky, 2019).

Within the agroecological sector, public authorities are believed to have contributed to the development of an agroindustrial model and to have seldom created policies seeking to address the inequalities and the socio-economic and environmental impacts of the food system (Espluga et al., 2017). Despite occasional transformative public policies, the majority of agroecological provisioning networks in Catalonia continue to struggle to safeguard the various aspects required to guarantee their sustainability.

[2] In accordance with European legislation approved in 1991, the Spanish Ministry transferred responsibility to the autonomous communities in 1994. Since then, the formal regulation system has been overseen by the Catalan Council for Organic Agricultural Production (CCPAE), which is a public law corporation with its own legal personality under the supervision of the Ministry of Agriculture, Livestock, Fisheries and Food of the Autonomous Government of Catalonia.

Economic Sustainability in Agroecological Provisioning Networks: Exploitation, Exclusion and Reproductive Work

Agroecological provisioning networks are based on relationships of trust, reciprocity and commitment between organised producers and consumers (Goodman, 2002). Their emergence has played a central role in the expansion of agroecology in Catalonia (López, 2015). On the one hand, they offer a concrete approach to building more sustainable food systems, which has helped support and consolidate agroecological initiatives. On the other, they represent a source of inspiration and promote agroecology across different layers of society (Michelini & Abad, 2017).

These networks usually take the form of organised collectives of consumers registered as an association and comprising 15–30 consumption units in direct contact with fresh food producers and processors. In most cases, food coops involve members only, have no shop to sell products to non-members and operate on the basis of members' unpaid work. Generally speaking, they do not require official organic certification and the relationship with farmers is based on mutual trust (Higgins et al., 2008). Many consumer cooperatives and consumer groups organise through working groups and decisions are made through assemblies. Meanwhile, production initiatives, organised into teams of members and salaried workers, cultivate between two and five hectares of vegetable gardens or manage small food processing facilities. Farmers supply consumer cooperatives directly without recourse to intermediaries.

The first consumer cooperatives were founded in the 1980s and have since proliferated. In 2016, there were around 160 cooperatives in Catalonia, although the census is incomplete because there is no up-to-date list of the number of initiatives (Fernández & Miró, 2016). However, growth has slowed in the sector since 2015, coinciding with the increased presence of organic products in conventional supermarkets (Martín et al., 2017). Besides structural factors associated with the broader socioeconomic context, this stagnation is also linked to internal limitations present in the most widespread model of agroecological networks in Catalonia. Some of the limitations identified in previous research include the economic unviability of farming initiatives; difficulties in the access to land and the means of production; sociological homogeneity in terms of class, gender, age, disability, dietary and culinary habits and cultural practices (Marfà, 2019; Ruiz Pascua, 2023); the failure to distribute care-related tasks within cooperatives (Rivera, 2018; Homs et al., 2021); dependence on unpaid labour (Homs et al., 2021); and the small size of the cooperatives (Martín et al., 2017). These limitations echo some of the challenges faced by alternative food networks that international scholars have identified, such as the lack of democratic access to agroecological provisioning systems; the lack of social justice; self-exploitation among agroecological farmers; the integration of alternative food networks in the capitalist economy; and the mechanisms by which these networks are scaled up and grow (Goodman & Goodman, 2009).

Against this backdrop, new cooperatives began to emerge in 2016 that sought to address the aforementioned challenges by creating organisational and operational structures on a larger scale, such as cooperative supermarkets with hundreds of members. This upscaling requires greater funding, the formalisation and professionalisation of the cooperatives and a greater degree of institutionalisation and proximity with the state (Homs et al., 2024; Homs, In Press).

In the production sphere, farmers report a series of obstacles to the economic viability of their initiatives, due in part to the amount of unpaid labour that they perform. For example, producers speak of poor economic planning, with initiatives failing to take into account the costs involved in maintaining a direct relationship between producers and consumers. Carmen, an agroecological vegetable farmer with more than 20 years' experience, explains:

> Yeah, we hadn't studied the production costs (...) we cut out the middlemen but we were the ones selling and distributing the products and that has a cost... you can't cut that out. But it's not real because we're doing the work, so we have to charge for it. That meant having a very low starting price, below cost, which allowed us to keep the project going, for the same reason as what's happening now, because there were a lot of hours of labour that were never paid and never will be.

Accounting for this labour in product pricing would mean increasing food prices and contributing to the exclusion of certain consumer groups from agroecological provisioning networks. There are considerable economic barriers determining who can and cannot participate in these networks. As a result, agroecological provisioning systems are characterised by constant tension between the desire to make them accessible to everyone and the exploitation present in labour relations, whether it takes the form of "self-exploitation" in the case of worker members or exploitation of salaried workers. Felipe, a salaried worker at a food cooperative, told us: "(…) we will never be viable. It's better to admit it early on. In any case, if we manage to be financially viable it will be when five or six people self-exploit and that's how the numbers work out". The idea that economic viability is impossible was repeated by several individuals involved in these networks, "in a market context we have everything to lose. We can't do anything. What differentiates us are our values". Echoing the studies of peasant farmers carried out by Chayanov (1966), highlighting the role of self-exploitation as a competitive strategy among small family farms, many of these networks do not account for the full costs of production and reproduction when pricing their products in order to remain competitive on the market.

These tensions affect the essential social and environmental values held by these networks (beyond the reproduction of capital), which can end up being overlooked: "The way things are, the first thing to do is to make it economically viable and we'll achieve our other social and environmental goals later", said Andrés, who works at a cooperative supermarket. They are also apparent in collective discussions concerning the criteria to be considered when deciding on the products that should be stocked by the supermarket. At a number of cooperative supermarkets, there have been debates around whether it is appropriate to stock conventional products, whose more affordable prices could boost sales and improve the cooperative's economic viability but jeopardise the social and environmental principles underpinning it.

Given these unsustainabilities within agroecological networks, farmers and consumers emphasise the need to cooperate with the public sector as a key strategy for improving their viability. To do this, they suggest interpreting (and justifying) subsidies[3] as a way of remunerating the unaccounted social and environmental values that have a positive impact on the community and the environment. In Felipe's words: "We have to change the idea of what a subsidy is. We have a social impact and that should be

[3] A complex institutional framework of subsidies encompasses municipal grants, subsidies issued by the autonomous communities or the State and European Union subsidies, which are usually handled by agencies in the autonomous communities. The largest and most well-known subsidies are issued under the Common Agricultural Policy (CAP) and have been criticised for promoting a specific model of production, which has sometimes prompted the modification of traditional, geographically situated forms of production (Escribano et al., Forthcoming).

paid for with public money". In other words, agroecological projects will not be economically viable and affordable for everyone in a capitalist market context: "we must be socioeconomically viable, not just viable in terms of the financial economy" and there must be "support from public institutions for the development and sustainability" of these initiatives. In this regard, specific proposals include linking public food procurement to agroecological networks; supplying canteens with agroecological products; connecting social policies such as cashless food cards and food banks with agroecological projects; making public infrastructure more accessible to all, for example municipal markets or logistical infrastructures such as Mercabarna; and compensating farmers for tasks that go beyond food production (preserving biodiversity, improving soil fertility, protecting from fires, etc.) and have a positive impact on the community and the environment.

Social Sustainability in Pastoral Systems in Catalonia: From Tension to Assimilation in the Market Economy

Shepherds are livestock farmers who specialise in harnessing variability, managing grazing routes at different scales to provide a better diet for livestock than it would obtain without the shepherd (Krätli & Nikolaus, 2010). There are many different types of shepherds (Scoones, 2020), just as there are different ways of moving animals to take advantage of pastures. Agropastoral systems have been conceptualised as allies of life, as they are associated with soil fertility, carbon storage and biodiversity creation (Montenegro de Wit, 2021, p. 115). Despite its positive impact on the environment, the profession struggles to ensure generational renewal (Badal et al., 2011). On the one hand, a decline in pastoral activity prompts the replacement of shepherds by immigrant workers from Albania, Romania, Morocco and Tunisia in particular (Scoones, 2020), modifying systems based on traditional family or community relationships. On the other, the sector is witnessing the arrival of people (both Spanish citizens and immigrants) who are aware of the problems facing pastoralism and of the importance of reviving it to counter the growth of agroindustry. A generation of new peasants and shepherds has emerged (Ploeg, 2008; Monllor, 2013), who were not necessarily born in rural locations and whose positionality could be described as "neo-rural" (Hummel &

Escribano, 2021), as it is defined not only by their geographical movement from the city to the countryside but also by their political stance of opposition to agroindustry. Among neo-rurals, there are also people who work as shepherds.

As a result of a growing critical awareness among pastoralists, whether or not they are neo-rurals, pastoral food systems have begun to adopt the goals of agroecology through production and consumption initiatives such as the provisioning networks explained in the previous case study. One example is Fernando, a 40-year-old goatherd with a flock of 600 animals in the Maresme area, who sells kid goats to his regional consumer group: "I don't do it for the money, I do it because I believe in it". Like other producers, Fernando negotiates the price for his animals with consumers, taking into account seasonal production needs. This arrangement gives him some leeway to mitigate the uncertainty of continuing production due to weather conditions and other factors. Despite such strategies, the sustainability of pastoral systems aligned with agroecology is constantly threatened by a variety of factors. One risk is that pastoralism may be stripped of its provisioning role, reduced instead to a leisure or folklore activity as cultural activities become more profitable than selling produce. Meanwhile, pastoralism faces pressures towards intensification in a quest to maximise profits, a process that undermines its holistic, organic character. In such cases, rather than serving as a potential solution, pastoralism risks becoming yet another environmental and economic problem, particularly due to soil and aquifer pollution caused by the stabling of animals.

Pastoralism in Catalonia evolved over thousands of years, as did the cultural and social norms governing the relationships between family farms, livestock farmers, pastoralists, transhumant communities, territories and systems of power. With the advance of industrialisation during the nineteenth century, the food provisioning system linked to livestock farming gradually shifted to become commodified within an agrifood industry. Social changes have been reflected in the legislation, but small farms, including agroecological farms, have fallen into a legal vacuum: without provision for such farms in the legal system, they are forced to comply with the legislation aimed at large farms, or else they have to operate a-legally or disappear. In turn, inflexible regulatory frameworks and the system of social and cultural incentives for industrialisation and profit maximisation have influenced pastoralist economic forms, bringing them closer to the institutional market model. Regulatory and market structures have thus compelled some farmers into intensifying production. Greater market orientation may lead to larger flocks. One example of the pressure to grow is the case of Pedro, who has 200 sheep and registered in a European Union funding programme for young farmers in 2018. He is 40 years old and lives with his wife and two children in the province of Girona. Confronted to bureaucratic pressures - such as an excessive burden of paperwork designed for large companies but unmanageable for a small family farm -, and economic pressures, particularly production costs leaving his business unviable, Pedro decided not to legalise his flock. Instead, he chose to focus on small-scale production for domestic use and on selling surplus produce. He plans to look for wage work in a different profession and has un-registered as a self-employed agricultural worker; he says with concern:

> More and more paperwork… and you end up trapped in a wheel that you have to make bigger and is always getting bigger. I've seen it among other shepherds too. The system obliges you to produce more and more, and the more you have the more they squeeze you and encourage you to get into debt. I want to be able to operate without getting into debt. But of course, 'you've got to modernise…' How many times have we heard that?! And who's paying for it?

The dismantling of small farms has been described as "depastoralisation", with reference to the "process of dispossession of the main means of social reproduction, which leads to an increase in livelihood diversification and social differentiation" within the world of pastoralism (Caravani, 2018, p. 1). When analysing the causes of this process of dispossession of the means of production in pastoralism, Spain's accession to the Common Agricultural Policy in 1986 emerges as a key factor in the evolution of food systems. Different rules and regulations intersecting at the European, national and autonomous community level have transferred a market-oriented food system model to the practice of pastoralism (among others), which has made social reproduction more difficult for small livestock farms, leaving them unable to compete and forcing many to shut down. It has become apparent that the CAP has left behind families who are not production-focused through the technification of the sector (Segrelles, 2017). In addition to the effects of the CAP, high land values and competition within the pig and sheep sectors are triggering a steady decline in the number of flocks in Catalonia: "There used to be two other flocks in this area… Now this is the last one, and I'm selling it". It is common to hear comments of this kind from agroecological pastoralists like David, a 48-year-old shepherd from the province of Barcelona. David had a 20-year rental contract for his farm, where he rears organic sheep and lives with his family. At the end of his contract, the owners of the farm put the property up for sale. This occurred against the backdrop of rising property values in rural areas during the COVID-19 pandemic. It was impossible for David to buy the farm, which was on sale for 1.2 million euros, or to start again at another property. He eventually decided to leave shepherding after 20 years at the farm and another 10 years as a salaried shepherd, and move to another area with his wife and three children. David fears that the farm will be purchased by a large-scale pig farming company, as the pig farming sector industry holds a large volume of capital in Catalonia: "There's no law to protect us, to protect all the work I've done at the farm [to convert it into an organic farm] and prevent it from being used to dump pig slurry". In the current economic model, a pig farm is more profitable than a flock of sheep. Indeed, the pig farming sector has generational renewal, new recruits and legislation suited to its activity.

Perceptions of the State in Relation to the Sustainability of Agroecological Networks

Amid the current context of ecosocial emergency, there is a growing awareness of the need for action at different levels to encourage ecosocial transition, from changing consumption habits to developing public policies that support sustainable food systems or large-scale cultural change (Herrero & Prats, 2024). In this regard, agroecological networks are a recognised, legitimate tool for making food systems more sustainable, fair and healthy. However, as the case studies show, agroecological provisioning networks are constrained by socioeconomic and cultural factors that hinder their growth and consolidation. The socioeconomic and cultural context in which agroecological initiatives operate makes social reproduction unviable. Against this backdrop, one option is to increase the amount of unpaid labour, or in other words, to seek competitive advantage in self-exploitation (Chayanov, 1966). The situation has increasingly tested the resilience of cooperatives, forcing some of them to consider abandoning or to abandon their initiatives. As an alternative to self-exploitation, agroecology has also explored partnerships with the state.

As we have seen, imaginaries of the state have evolved over the past 30 years. In the late 1990s and early 2000s, agroecological networks tended to reject the intervention of the state, viewing it as incompatible with their "*self-management and autonomy*", "*the transformation of food systems towards sustainability and fairness*" and "*bottom-up processes*" (field notes). This rejection was partly due to the state's role as a key actor in the reproduction of capital by agroindustry and it manifested itself among agroecological initiatives in ways such as avoiding formalisation (Escribano, 2023). For example many organised consumer groups were a-legal in the sense that they were neither registered as associations nor as consumer cooperatives (Homs & Narotzky, 2019). Nevertheless, over the years, many have officially registered as consumers' cooperatives and this has brought about a shift in their self-perception:

> Another thing that happened when we legalised our activity as a cooperative is that we realised that we had to be a company. That's something that nobody talks about because everyone's kind of ashamed, everyone avoids talking about companies. It was a heated debate... We wanted to be a different kind of company, but we had to live in a context with an economic system that we're all too familiar with.

In (semi-)extensive livestock farming, the registration of animals for health checks occurs in parallel with the acquisition of legal status by agroecological initiatives; this means that some farms may have registered their animals but not their commercial activity, due in some cases to the costs involved.[4] In order to officially register the sale of their products, they must comply with current regulations that are not designed for small-scale farming. People in the sector are aware that the export of pig meat from Catalonia, which is the main economic driver of the food system, affects the administrative procedures to be carried out by pastoralists. Fears among public authorities over the spread of animal diseases (such as bluetongue or brucellosis), which can have a major impact on the economy, prompt them to impose the same requirements on all farms, regardless of their size. "They're asking me to set up an area to disinfect myself every time I go in and out of the yard. But my daughters are always running

[4] In Spain, the average monthly Social Security payment for self-employed people is around €300. This payment varies according to the chosen contribution base and the type of contribution that applies. There may be different types of contribution depending on the applicable scheme, for example the Special Scheme for Self-Employed Agricultural Workers.

around out there and I'm constantly going between the house and the yard. It's impossible for me to set up a system like that here on the farm", says Carla, a 38-year-old farmer with a flock of 100 goats and (a)legal artisan cheese factory on her farm.

Another example of the shift from rejection to acceptance of state regulations within agroecological networks is their position regarding theofficial organic certification. The predominant discourse tends not to value organic certification because it is a "*top-down*" mechanism that obliges farmers who "*do things properly*" to pay, while promoting a standardised idea of what "organic" means and even appropriating the concept. A member of a consumer group explains: "there's no need for the seal because it belongs to the authorities, it's part of the system, and it's always the same people taking the money". However, in agroecological networks based on agriculture and livestock farming, the organic seal has become increasingly present in short circuits, as it is seen as a strategic tool for increasing sales channels aimed at larger sectors of society, such as school canteens, small agroecological shops and cooperative supermarkets.

Amid the shift from rejection to recognition of the key role played by the public sphere in the development and consolidation of agroecological provisioning networks, internal tensions are emerging within these networks as people raise the alarm over the possible depolitisation processes, while acknowledging the need for a change of model in order to enable the social reproduction of agroecological networks. As Carmen, an organic vegetable farmer, told us:

> It's like 'am I selling out?', right? You're selling yourself to capitalism. Like, we're looking for alternatives to the system, but we're part of it. So, for us to be viable within this system, we're moving closer and closer to strategies that are already in place within the system. So, it's like… it makes you doubt whether you'll be able to maintain your integrity or that basic essence I mentioned. I mean, fuck, there comes a point where I don't want to lower my expectations any further (…) So… either we change or we might end up disappearing.

Despite the inherent contradiction in accepting support from the state, which is associated with support for agroindustry, many agroecological networks view this as their only option for survival at present.

The change in perception between agroecology and the state has not only occurred from the perspective of agroecological networks. Since

2020, institutional recognition of agroecology as a tool for ecosocial transition in Catalonia has grown in a process that has also been observed in other regions (Giraldo & McCune, 2019; Giraldo & Rosset, 2017). This recognition is largely limited to agroecological initiatives that are formalised and comply with all current legislation, which sometimes entail less transformative agroecology practices and values. As a result of this institutional recognition, a variety of municipal and regional food strategies have been put in place, prioritising the need to move towards a more sustainable and healthy food system for people and the planet. In this context, agroecology has gained legitimacy in the social imaginary as a step towards this goal. Specific examples include the Strategic Food Plan for Catalonia (PEAC), the Barcelona Healthy and Sustainable Food Strategy and food strategies for other locations, as well as grants awarded to cooperative supermarkets or other agroecological initiatives.

The shift in perceptions of agroecology within the public institutions is also apparent in the emergence and gradual increase in the hiring of small flocks or herds of animals to clear forest belts. Fernando described his involvement in this activity in 2021: "Now we take the herd to clear the areas that we're asked to clear. They can pay us up to €10,000 per year for this work. When they came to see us to check that we were suitable, they told us: 'this is exactly what we're looking for!'". Currently, he is earning money for clearing vegetation in countryside areas in the province of Barcelona with his herd of 600 goats. Fernando was surprised to see this shift in his relationship with the authorities: "Initially, nobody wanted us. They'd look down on us. It was almost like we were terrorists. Now, in some places, we're practically seen as heroes". The improved social status of (semi-)extensive pastoralists during the COVID-19 pandemic also contributed to this change in perception among the authorities: "I'd never felt like this before, they were treating us like heroes", said David when asked about the pandemic. "During the initial opening up after the lockdown, people would approach us and ask to walk with us. Some of them even bought a goat off me", said Fernando. Both men agreed that their popularity had begun to decline again when "normal life" returned, although this period of improved social status contributed to a shift in perception within public institutions. Fernando even began to notice that his administrative procedures were handled more quickly by the local council.

As recognition from the state has shifted, so has recognition and legitimation of the public sphere within grassroots agroecological movements, which have gone as far as to frame the participation of the state in terms of necessity, requirement or even imperative: "*they must pay us for what we do*". It is now common for agroecological cooperatives to advocate public support for their initiatives and raise the possibility of building public-cooperative-community alliances to strengthen the bonds between the cooperative, community and public spheres and enhance the sustainability of agroecological projects. They call for public policies to facilitate the growth of agroecology and to encourage replication, cooperation and interconnection between initiatives, which play a crucial role in scaling up agroecology (López et al., 2017; Sanz-Cañada et al., 2023). In this regard, they also draw attention to the need to "go further than sporadic support for initiatives in the form of one-off subsidies and promote structural changes to public policy" that improve (and guarantee) the sustainability of these types of initiatives. Thus, the participation of the state and the development of policies that reflect the needs of agroecology are considered essential to safeguard the various dimensions of sustainability present in agroecological provisioning networks.

Navigating Opposition and Recognition: Agroecology and the State

Agroecology as a concept emerged out of concern among farmers, activists and academics over the need to collectively build food systems that would counter the extractivist logics of intensive industrial agriculture. As we have seen, agroecology has spread across Catalonia and Spain with the goal of transforming market-oriented forms of production, distribution and consumption, and of supporting the subsistence economies and traditional knowledge that continue to exist in rural cultures. Through our research, we have analysed how agroecological provisioning networks based on the reproduction of life are economically unviable, hindering their social reproduction, in a market and institutional context in which regulations adapted to agroecological or small-scale farming practices are lacking. Practices that are aimed not at accumulating capital but at sustaining the lives of people, the territory and the planet have led to the reproduction of vulnerabilities that have been exacerbated by recurring capitalist crises. Among these vulnerabilities, we have identified (1) a growing tension due to the rise of (self-)exploitation in response to competition on the capitalist market; (2) lack of recognition and appreciation of agroecological practices from society and public institutions; and (3) exhaustion and burnout among agroecological farmers due to their constant struggle against the hegemonic system.

In response to ongoing vulnerabilities and difficulties with regard to social reproduction, agroecological initiatives have been forced to rethink their discourses, values and practices. In some cases, they have adopted a different strategy by setting higher prices, although this makes their products less affordable for all social groups. In other cases, they have opted to stock non-agroecological products in cooperative supermarkets to boost sales and enhance the economic viability of these initiatives, with the risk of reproducing forms of consumption that follow the same extractivist logics that they had sought to avoid. Other strategies include increasing the number of animals kept and applying more intensive techniques that make farms more profitable. Those who do not wish to adopt any of these strategies have abandoned their initiatives and/or sold their animals.

Along the last 30 years, the relationship between the agroecological movement and the state has gradually changed. In response to greater openness from public authorities, members of agroecological provisioning networks have called for the state to intervene to guarantee their sustainability, or even demanded such intervention as a social debt. Not only do they demand more direct economic support in the form of subsidies for their projects, they also advocate structural changes to food policies to ensure the viability and sustainability of these networks from a more systemic, long-term perspective (Espluga et al., 2017; Calle Collado et al., 2021).

Despite explicit warnings that grassroots agroecological movements could be coopted and depoliticised, we are witnessing a shift in the positionality of agroecology in relation to the state. While agroecology embraces and, in some cases, demands public support, there is a need to reflect on the tensions between the different logics present within the "alliance" among agroecology and the state. It would be useful to continue to analyse these tensions and the effects of changing relationships and imaginaries between the state and agroecology on community bonds, social organisation and collective decision-making processes, among other areas.

References

Badal, M., Binimelis, R., Gamboa, G., Heras, M., & Tendero, G. (2011). *Arran de terra: indicadors participatius de sobirania alimentària a Catalunya. Informe*. Associació Entrepobles and Institut d'Economia Ecològica i Ecologia Política. https://entrepueblos.org/wp-content/uploads/1970/01/informeindicadorssba1-catalunya.pdf

Batalla, L., Correro, A., Hierro, A., Homs, P., Ravera, F., Rivera, M. G., & Sastre, A. (2020). *Xarxes alimentàries locals en temps de COVID-19*. https://arrandeterra.org/wp-content/uploads/2020/07/Xarxes-aliment%C3%A0ries-locals-COVID19_DEF.pdf

Calle Collado, A., Gallar, D., & Candón, J. (2021). Agroecología política: transición social hacia sistemas agroalimentarios. *Revista de Economía Crítica, 2*(16), 244–277.

Caravani, M. (2018). 'De-Pastoralisation' in Uganda's Northeast: From Livelihoods Diversification to Social Differentiation. *Journal of Peasant Studies, 46*(7), 1323–1346. https://doi.org/10.1080/03066150.2018.1517118

Chayanov, A. V. (1966). *The theory of peasant economy*. American Economic Association.

Cuéllar, M., & Sevilla, E. (2013). La soberanía alimentaria: la dimensión política de la agroecología. In M. Cuéllar, A. Calle, & D. Gallar (Eds.), *Procesos hacia la soberanía alimentaria* (pp. 15–32). Icaria.

Escribano, P. (2023). Playing by the rules: Formalisation in the agroecology sector during the COVID-19 pandemic. *Journal of Peasant Studies, 51*(6), 1428–1451. https://doi.org/10.1080/03066150.2023.2285478

Escribano, P., Hummel, A., & Madriles, D. (Forthcoming). Your herd or your life! Long-term de-pastoralization and public policy in Catalonia, Spain. *Focaal: Journal of Global and Historical Anthropology*.

Espluga, J., Di Masso, M., & López, D. (2017). Agroecologia i polítiques per la sobirania alimentaria. In R. Gomà & J. Subirats (Eds.), *Canvi d'època i de polítiques públiques a Catalunya* (pp. 236–242). Crític SCCL.

Fernández, A., & Miró, I. (2016). *L'economia social i solidària a Barcelona*. La Ciutat Invisible SCCL and Comissionat d'Economia Cooperativa, Social i Solidària – Ajuntament de Barcelona.

Fraser, N. (2013). *Fortunes of Feminism: From State-Managed Capitalism to Neoliberal Crisis*. Verso Books.

Friedmann, H. (2005). From colonialism to green capitalism: Social movements and the emergence of food regimes. In F. H. Buttel & P. D. McMichael (Eds.), *New directions in the sociology of international development. Research in rural sociology and development* (Vol. 11, pp. 227–264). Elsevier.

Gascón, J. (2020). COVID-19, state of emergency and family farming in Spain: Rural markets in Barcelona and self-consumption orchards in Alcaine (Teruel).

Ager. Revista de Estudios sobre Despoblación y Desarrollo Rural, 30, 177–206. https://doi.org/10.4422/ager.2020.13

Giraldo, O. F., & McCune, N. (2019). Can the state take agroecology to scale? Public policy experiences in agroecological territorialization from Latin America. *Agroecology and Sustainable Food Systems, 43*(7–8), 785–809. https://doi.org/10.1080/21683565.2019.1585402

Giraldo, O. F., & Rosset, P. (2017). Agroecology as a territory in dispute: Between institutionality and social movements. *Journal of Peasant Studies, 45*(3), 545–564. https://doi.org/10.1080/03066150.2017.1353496

Goodman, D. (2002). Rethinking food production-consumption: Integrative perspectives. *Sociologia Ruralis, 42*(4), 271–277.

Goodman, D., & Goodman, M. K. (2009). Alternative food networks. In R. Kitchin & N. Thrift (Eds.), *International encyclopedia of human geography* (pp. 208–220). Elsevier.

Graeber, D. (2011). *Debt: The first 5,000 years.* Melville House.

Guzmán, G., González de Molina, M., & Sevilla, E. (2000). *Introducción a la agroecología como desarrollo rural sostenible.* Mundi-Prensa.

Harvey, D. (2011). *The enigma of capital and the crises of capitalism.* Profile Books.

Herrero, Y., & Prats, F. (2024). La transición ecosocial justa en España. *Sistema: Revista de Ciencias Sociales, 269–270*, 179–194.

Higgins, V., Dibden, J., & Cocklin, C. (2008). Building alternative agri-food networks: Certification, embeddedness and agri-environmental governance. *Journal of Rural Studies, 24*(1), 15–27. https://doi.org/10.1016/j.jrurstud.2007.06.002

Homs, P. (In Press). Alimentar la vida: tensiones y desafíos en el cambio de escala del cooperativismo agroecológico en Catalunya. In S. Sama Acedo & J. Sanz Abad (Eds.), *Dilemas y desafíos de las economías transformadoras en un contexto de urgencia ecosocial.* Editorial Bellaterra.

Homs, P., Flores-Pons, G., & Martín Mayor, A. (2021). Sustainable lives and care: agroecological cooperativism in Catalonia, Spain. In A. Nelson & F. Edwards (Eds.), *Food for degrowth (pp. 90–111).* Routledge.

Homs, P., & Narotzky, S. (2019). Within and beyond the market system: A case of organic food cooperatives in Catalonia. In V. Siniscalchi & K. Harper (Eds.), *Food values in Europe: Economies, ideologies, and power in practice* (pp. 132–146). Bloomsbury.

Homs, P., Sama, S., & Berná, D. (2024). La sostenibilidad del cooperativismo agroecológico: desafíos de los supermercados cooperativos y otros modelos de crecimiento en Cataluña. *REVESCO: Revista de Estudios Cooperativos, 149*, 1–14.

Hummel, A., & Escribano, P. 2021. The neo-peasant movement in Catalonia: an attempt of defining it in the light of the infrapolitical strategies of resistance. *Sociologia Ruralis, 62*(1), 3–23. https://doi.org/10.1111/soru.12255

Idescat. (2024). *Estadística de la ramaderia*. Retrieved April 6, 2025, from https://www.idescat.cat/pub/?id=proram&utm_campaign=cercador&utm_medium=sugg&utm_source=Idescat&utm_term=ramaderia&utm_content=proram

Krätli, S., & Nikolaus, S. (2010). Living off uncertainty: The intelligent animal production of dryland pastoralists. *European Journal of Development Research, 22*(5), 605–622. https://doi.org/10.1057/ejdr.2010.41

López, D. (2015). *Producir alimentos, reproducir comunidad: Redes alimentarias alternativas como formas económicas para la transformación social y ecológica*. Ecologistas en Acción.

López, D., Fernández-Casadevante, J. L., Morán, N., & Oteros, E. (2017). *Arraigar las instituciones: Propuestas de políticas agroecológicas desde los movimientos sociales*. Ecologistas en Acción.

Marfà, M. (2019). *La inclusió en el cooperativisme de consum agroecològic a Catalunya* [Master's dissertation, IGOP-UAB, UdL, UVIC & UdG].

Martín, A., Homs, P., & Flores-Pons, G. (2017). *El canvi d'escala: un revulsiu per a la sostenibilitat del cooperativisme agroecològic?* L'Aresta Cooperativa Agroecològica.

McMichael, P. (2009). A food regime genealogy. *Journal of Peasant Studies, 36*(1), 139–169. https://doi.org/10.1080/03066150902820354

Michelini, J., & Abad, L. (2017). Nuevas prácticas colaborativas, cuestión alimentaria y desarrollo urbano: Los grupos de consumo agroecológico en Vallecas, Madrid. In J. Delgadillo & J. Sanz (Eds.), *Sistemas agroalimentarios locales de proximidad en contextos rururbanos en México y España* (pp. 73–101). Universidad Nacional Autónoma de México.

Montenegro de Wit, M. (2021). What grows from a pandemic? Toward an abolitionist agroecology. *Journal of Peasant Studies, 48*(1), 99–136. https://doi.org/10.1080/03066150.2020.1854741

Oñate, J. J., Acebes, P., & Olea, P. P. (2023). Aprender del pasado para afrontar el futuro. Desafíos ambientales de la agricultura española en el siglo XXI: una mirada desde el legado de Fernando González Bernáldez. *Ecosistemas, 32*, 2495. https://doi.org/10.7818/ecos.2495

Pomar, A., Durant, N., Gamboa, G., Binimelis, R., & Tendero, G. (2018). *Arran de terra II. Indicadors de Sobirania Alimentària a Catalunya*. Associació Entrepobles & Arran de Terra. https://web.archive.org/web/20220820014808/http://indicadors.arrandeterra.org/informe/

Rivera, M. (2018). L'agroecologia és feminista? Algunes reflexions desde la teoria i la pràctica. *Agrocultura, 73*, 6–8.

Ruiz Pascua, M. (2023). *Pel Dret a una alimentació culturalment adequada: diagnosi etnogràfica i propostes agroecològiques*. L'Etnogràfica - Antropologia per a la Transformació Social.

Sanz-Cañada, J., Sánchez-Hernández, J. L., & López-García, D. (2023). Reflecting on the concept of local agroecological food systems. *Land, 12*(6), 1147. https://doi.org/10.3390/land12061147

Scoones, I. (2020). Pastoralists and peasants: Perspectives on agrarian change. *Journal of Peasant Studies, 48*(1), 1–47. https://doi.org/10.1080/03066150.2020.1802249

Segrelles, J. A. (2017). Las ayudas agrarias y sus repercusiones sobre la agricultura familiar en la última reforma de la política agraria común (2014–2020) de la unión europea: ¿cambiar todo para que todo siga igual? *Boletín de la Asociación de Geógrafos Españoles, 74*, 161–183. https://doi.org/10.21138/bage.2449

Sevilla-Guzmán, E. (2011). *Sobre los orígenes de la agroecología en el pensamiento marxista y libertario*. Plural Editores.

Sevilla-Guzmán, E., & Woodgate, G. (2013). Agroecología: fundamentos del pensamiento social agrario y teoría sociológica. *Agroecología, 8*(2), 27–34.

Shiva, V. (2016). *¿Quién alimenta realmente al mundo?* Capitán Swing.

van der Ploeg, J. D. (2008). *Los nuevos campesinos: Campesinos e imperios alimentarios*. Icaria Editorial.

Cultivating Alternative Supply Chains: Hemp Gardens in a (Post)industrial Valley

Davide Cacchioni

In this chapter, I examine three gardens where hemp is cultivated in the Susa Valley, an alpine region in Northwestern Italy situated between Turin and the French border. Here hemp farming is part of a project of "hemp supply chains", explicitly mobilized to imagine a different economic organization, based on hemp transformation across multiple supply chains. Hemp supply chains are thought of as capable to support an economy that can meet human needs by sourcing plant resources from non-extractive ecologies. I describe the gardens where hemp is grown as "niches", borrowing Anna Tsing's conceptualization of "supply chain capitalism" (Tsing, 2009). For Tsing a niche is a cultural, economic, and ecological context connected to capitalist chains (Tsing, 2009, 2012, 2015). I apply this framework to the "hemp supply chains" I encountered in the valley to describe the ecological and economic context of hemp farming, and to extend my focus to processing initiatives. Following a certain tradition in the anthropology of Alpine spaces (Cole & Wolf, 1974/1999; Rosenberg, 1988), observing hemp initiatives has led me to focus on the different

D. Cacchioni (✉)
EHESS-CeRCLEs, Marseille, France

S. Narotzky et al. (eds.), *Agricultural Extractivism in the Mediterranean Region*,
https://doi.org/10.1007/978-3-032-05599-6_12

political-economic and political-ecological projects existing within the same valley, the actors involved in the hemp project, its characteristics, difficulties, and failures. I conducted 15 months of fieldwork in the Susa valley, between 2015 and 2019, based on participant observation and ethnographic interviewing. The research included various actors, ranging from hemp farmers to individuals involved in processing initiatives and local administrators. The case of "hemp supply chains" prompted me to shift the focus on the Susa valley—which has experienced a long-lasting mobilization against a large-scale infrastructure project—from the "struggle for subjectification" (Leonardi, 2013) to the emergence of alternative economy initiatives. I argue that an ethnographically grounded exploration of hemp farming and processing in this region can contribute to the anthropological scholarship on alternative supply chains by focusing on elements that prove to be central to these economic experiences: capital and technology, the relationship with the dominant economy, and ideology forms.

The "hemp supply chains" in Context: The Susa Valley Between No TAV Mobilization and Economic Initiatives

The Susa Valley (*Val di Susa* in Italian) is an alpine territory, inhabited by several tens of thousands of people, between the urban and (post)industrial city of Turin, in Northwestern Italy, and the French border. Two mountain passes, the Col du Mont-Cenis and the Col de Montgenèvre, connect the valley to the French Alps. From the late nineteenth century onwards, the Susa Valley has undergone significant industrial development with various industries—electrical, textile, iron and steel, mechanical, chemical industries, and railway workshops—spreading throughout the area, particularly close to the Turin industrial hub (Sacco, 2020). In the last decades of the twentieth century, the model of the large-scale enterprise entered a major crisis in Northwestern Italy, epitomized by the downsizing of one of its most emblematic representatives, FIAT[1] (Berta, 2006). In this context the valley experienced a series of plant closures,

[1] The number of manufacturing employees in the north-west of Italy fell from over 2,512,000 in 1980 to 1,712,000 employees in 1996. FIAT (*Fabbrica Italiana Automobili Torino*), the largest automobile manufacturer in Italy, more than halved its workforce in the north-west between 1980 and 1996: from 261,000 employees to 127,600 (Berta, 2006,

reducing the scale of industrial production. Alongside the class-based demands raised during workers' struggles for better working conditions and in defense of industrial employment in the valley,[2] new critiques of development also began to emerge on nonviolent[3] and environmental bases. In particular between the 1980s and the 1990s, environmentalist mobilizations challenged major infrastructure projects, such as the Turin-Bardonecchia highway and a new power line. The emergence of the No TAV movement in the 1990s and 2000s marked a qualitatively significant shift, as it came to represent a broader and more culturally heterogeneous mobilization (Caruso, 2010; Aime, 2016).

The No TAV movement contests the construction of a new high-speed/high-capacity railway line between Turin and Lyon. The so called "TAV" (*treno ad alta velocità* in Italian) project was initially proposed in the late 1980s and early 1990s, and has, since then, been revised multiple times (Pepino & Revelli, 2012). Opposition to this infrastructural project is grounded in economic, transport-related, democratic, and environmental concerns. Critics question the need for a new railway line, given the presence of an existing international railway running through the valley. They also denounce the disregard for the local opposition and criticize the ways the State and public authorities manage dissent (Chiaramonte, 2019). Furthermore, the movement emphasizes the project's detrimental effects on the valley's environment and public health (Laszczkowski, 2019), as well as the potential for criminal infiltration in the construction sites.

The conflict over the proposed new railway line between Turin and Lyon in the Susa valley has contributed to the politicization of the material flows underpinning the economy. From the perspective of its critics, the "megaproject" (Laszczkowski, 2025), designed to transport people and goods, is being imposed on a territory already subject to intense anthropogenic pressures. The Susa Valley is currently traversed by two national roads, a highway, and an existing railway line, while the valley floor is

pp. 236–237). In the Susa Valley several historical industrial activities came to an end in the same period.

[2] For example, in the mobilization in the *Cotonifici Valle Susa* in the early 1960s (Accornero, 2011) or during the struggles that concerned the ASSA steel mill in Susa in the 1970s (Sacco, 2002).

[3] As a case in point, on 24 September 1970, the assembly of around 800 workers at the *Moncenisio* plant in Condove expressed the unanimous opposition to any conversion of production to military purposes (Oliva, 2020, pp. 38–42).

occupied by industrial facilities and residential areas. As such, opposition to the TAV project entails a broader critique of the material flows sustaining its economic rationale. This interview excerpt with Luciano, a No TAV activist, elucidates the point:

> [We began to] realize that [our] everyday choices either support or challenge the implementation of freight-traffic systems. If we demand goods that have traveled long distance, we're, in a sense, asking for the high-speed train that moves them. If we [...] want to oppose that, we need to start thinking about how to undermine this mechanism at its root, don't we? So, what can I do in my daily life, in addition to demonstrations? (Luciano[4], interview, Susa valley, November 2015)

The "development model" embodied by large-scale infrastructure projects is denounced by No TAV activists while "degrowth" perspectives gain traction. In particular, the approaches advocated by Serge Latouche (Latouche, 2007), who visited the Susa Valley in 2006, have found resonance within the movement. In this sociopolitical and ideological context, everyday economic practices come into focus as a site of engagement, and a political connection is established between ordinary life and broader systemic change. For instance, the choice to consume locally produced goods acquires political significance, representing a commitment to an economy that renders transport infrastructure, such as the new Turin-Lyon railway, redundant. Solidarity Purchase Groups (Grasseni, 2014; Dubuisson-Quellier, 2018) have proliferated across the valley, based on citizens' collectives that circumvent large retailers and prioritize local producers for sourcing both food and manufactured goods. The critique of the TAV project and the implications it makes evident in people's daily lives translate into efforts to reorganize the economy at a local scale. These initiatives extend beyond individual or household consumption to include productive activities, echoing dynamics observed in other contexts (Schlosberg & Coles, 2016). For several years, an association of so-called "ethical entrepreneurs", *Etinomia*, has sought to bring together local business owners who reject participation in the "large-scale projects" economy and instead advocate for "small-scale projects". As one member explained, "what we wanted to say is that it is not necessary to destroy a

[4] The names of people interviewed and cited in the ethnographic sections are anonymized. The texts quoted have been translated by the author.

valley or to make an economy that is not honest in order to provide jobs and to make a profit" (*Etinomia* member cited in Aime, 2016, p. 223). In the agricultural sector "alternative" initiatives have been also developed, including cooperatives and associations established to support local production, employing ecological farming practices (Caruso, 2010; Aime, 2016; Cacchioni, 2021a). In what follows, I focus on the "hemp supply chains" project, which emerged in 2010s within this ideological and practical context of economic experimentation.

In the Susa valley, hemp is promoted as a culture and an economic resource by an association, which I call *Hemp&Hemp*. Once a historic crop in Italian agriculture and industry, hemp cultivation virtually disappeared from post-war Italy. Its long, laborious, and labor-intensive transformation process was rapidly displaced by alternative materials, like synthetic fibers, in application ranging from textile production to maritime equipment (Poni & Fronzoni, 2005b; Cacchioni, 2021b). After several decades in which hemp "[had] almost lost as a word [...] the meaning of textile plant and fiber" (Poni & Fronzoni 2005a, p. VII), the plant began to attract renewed attention in the late twentieth century. It came to be reimagined as a natural fiber with the potential to re-localize fiber production in Italy and reduce the reliance on synthetic materials in the national apparel industry. Despite this revival, initiatives to reintroduce hemp cultivation and processing in Italy have met with limited success.

For *Hemp&Hemp*, the initial objective was to reintroduce hemp cultivation in the Susa valley, where its disappearance mirrored the decline of hemp industry at the national level. Once reintegrated into the local agriculture, hemp was envisioned as a resource that could develop multiple local supply chains (*filiere* in Italian). A highly "versatile" crop (Poni & Fronzoni, 2005b), hemp can be used for a variety of applications beyond textiles. In the Susa valley, the main areas of experimentation have included sustainable building materials and food production, and textile manufacturing. These "hemp supply chains" are conceived as a grassroots response to the context of economic crisis experienced by the valley's population (Narotzky & Besnier, 2014; Narotzky, 2020):

> [...] In the territory of the Susa Valley we are witnessing a progressive impoverishment caused by the economic and industrial crisis. Far from ending, the crisis, continues to result in job losses. [...] [In response] we asked ourselves how we might address people's needs by finding raw materials in the territory and asking the territory itself to produce and process them. We

> asked ourselves: 'what is needed?' Although the needs are many and the reflection could actually become immense, our aim is to offer a modest response through the development and, hopefully, the realization of our various supply chain projects. (*Hemp&Hemp* presentation document)

The multiple possibilities of transforming hemp into clothing, food products, or building materials made it a promising response to different local "needs". In a valley affected by crisis, these applications offered the potential to generate new productive activities and employment opportunities. Linking agriculture and industry, the "hemp supply chains" were also conceived as an experiment with a different production model: one rooted in the local provisioning of raw materials and the development of a small-scale, "ethical" processing industry (as described by project's participants). This model stood in stark contrast to the extractive and expansive logic of infrastructural projects.

Since the 2010s, the association has sought to develop these supply chains within the limits imposed by its financial means, reflecting the characteristics of a collective of a few dozen individuals from working- and middle-class backgrounds, employed in services, artisanal trades, and small-scale farming. The structure of the association did not permit the development of economic projects reliant on the mobilization of external capital. Collaboration between small-scale farmers and artisans (Cacchioni, 2021a) has therefore largely replaced joint investments. The "hemp supply chains" project has remained very marginal in terms of its productive outcomes. Nevertheless, it has succeeded in building a network of people mobilized around hemp cultivation, achieved limited production within the food supply chain, and articulated a broader economic imagination. I will further illustrate these dimensions in the following ethnographic vignettes.

Three Hemp Gardens as Economic and Ecological Niches

Hemp cultivation in the Susa Valley has developed through small- and micro-scale agricultural initiatives, scattered across the lower and upper valley and its surrounding territories. In the following section, I present three cases of cultivation (or, in the second case, non-cultivation) that highlight key aspects of the Susa valley context and the "hemp supply chains" project. Tracing hemp cultivation across the valley reveals differences among the niches where the plant is grown. As shown in the ethnographic vignettes that follow, hemp traverses multiple ecologies. First, we encounter the upper valley, where the environmental conditions and climatic constraints shape both the rhythm and the context of agricultural life. Moving down, we find the peri-industrial zones surrounding several small towns in the lower valley. Finally, we arrive at one of Giovanni's scattered plots, part of a constellation of small-scale farmlands. These vignettes will show how actors involved in the "hemp supply chains" project have engaged in productive experiences.

Garden One: "We are small, distant islands"

Lucia and Enrico grow hemp in the upper Susa Valley, at an altitude of over 1000 meters. Their garden—one of the highest in the valley—is situated in a municipality that belongs to a vast ski area in the upper Susa Valley. This ski resort includes about 400 km of slopes ranging between 1350 m and 2800 m in altitude and was one of the venues for the 2006 Winter Olympics. The local economy is heavily oriented around tourism: many of Lucia and Enrico's neighbors are employed in the ski lift operations or in sectors tied to seasonal hospitality and related activities. The economy of the villages in the upper Susa Valley is shaped by the demands of tourism, including the construction of accommodation facilities and second homes (Bermond, 2007), and by the rhythms of service work associated with ski-related and seasonal tourism.

Lucia and Enrico run a small guesthouse where they host tourists overnight and offer breakfast and dinner. They deliberately choose not to serve lunch, worried that it would turn the establishment into a destination for day-trippers arriving by car to eat. When they took over the facility, their goal was to contribute to a form of low-impact tourism oriented towards those interested in exploring the surrounding mountains on foot, whether in summer or winter. This kind of tourism stands in stark contrast to ski lifts, helicopter skiing and other motorized mountain activities dominating the region's economy.

When we met in 2018, Lucia and Enrico were in their late 20s to early 30s. They were not born or raised in the village where they now live, although Enrico had visited it sometimes during school holidays as a child, following his family's fascination with the mountains. Both Lucia and Enrico originally come from the Po Valley and have spent time in urban settings. Their decision to move up into the mountains carried, for them, a strong ethical and political significance: an expression of simplicity and withdrawal from the logic of work and production, as Lucia describes it.

During the short growing season allowed by the mountain climate, Lucia and Enrico tend various crops. In the garden next to their house, they grow vegetables for their guesthouse kitchen, and aromatic herbs such as thyme, marigold, mint, and mallow. On a small plot of land slightly removed from the guesthouse they grow hemp. However, the altitude of the garden prevented the seeds of the plant from ripening and the garden has not yielded any significant production yet. Because the hemp seeds are the basis of the food supply chain, the inability to harvest ripe seeds prevented them from taking part in *Hemp & Hemp*'s food production activities.

Why do they continue to grow hemp even if it brings no economic return? Lucia explains that growing hemp requires little effort, yet it keeps them connected with others involved in the association's projects, such as food or bio-building materials supply chain initiatives, which I will observe in the following gardens. Hemp links them to people "you feel are more like you" as Lucia put it during an interview (September 2018). Sometimes they drive more than 60 km each way to attend association meetings. The other farmers in the association are a point of reference for them, not only for agriculture advice, but also for broader exchanges of ideas. "We are small, distant islands", Lucia resumes, implying that hemp has created opportunities for connecting with others beyond their village: these hemp connections make them feel less alone.

Garden Two: Phytoremediating? A Missed Garden

On a winter evening in 2019, a gathering was held in a small municipality in the lower Susa valley to discuss the pollution legacy left by decades of industrial activities in the lower valley. A university expert presented findings about various contaminants detected in the soils of the host municipality and neighboring villages. These included micropollutants, hazardous substances capable of causing harm even at very low concentrations. Tracing and quantifying their presence is technically complex and costly. That evening, the discussion primarily focused on dioxins and PCBs: how they are produced and released into the environment, and how humans and animals are exposed to them. These molecules are mainly found in the soil, where they risk being transmitted to human and animal food. The expert described an "overall cultural transformation" triggered by pollution in some areas of the valley: agriculture and livestock farming had become unviable due to the presence of these substances in the local environment. However, since the 2000s, the pressure from the public and the institutions has increased after high pollution levels were discovered in 2004. Emissions were reduced and eventually halted, thanks to new regulations and, later, the shutdown of some industrial activities in the 2010s. Yet the soil in parts of the valley remains polluted, particularly at certain depths.

The meeting that evening—co-organized by the *Hemp&Hemp* association and the local municipality—was driven by a specific proposition: that hemp could be cultivated on polluted soils and repair the damage, taking advantage of its phytoremediation properties. These qualities are relatively well known among those familiar with the plant. In Italy, for example, an initiative to use hemp to remediate contaminated soil in Taranto—a city in the south heavily impacted by pollution from a large steel plant—had already drawn public and media attention (Martini, 2016).

If tests on the harvested plants confirmed they were non-toxic, the grown hemp could then be used in a sustainable building material's supply chain. Abandoned sheds from former industrial enterprises in nearby villages could serve as potential space for collecting and processing the harvest. The hemp could subsequently be transformed into bricks or other building materials. In this way, hemp cultivation was imagined as serving multiple purposes: contributing to soil remediation, enabling the re-use of dismissed industrial infrastructures, and reorienting them towards new forms of small-scale industrial production. This was the broader vision that the *Hemp&Hemp* association, together with local interested municipalities, began to develop over the course of several meetings during the winter of 2019.

However, this collaboration and its underlying vision never materialized into a full-fledged project. Multiple obstacles hindered its realization: the high costs of testing the harvested plants, the risks associated with handling polluted soils, and the uncertain efficacy of hemp alone in remediating pollution posed significant challenges to the project's implementation. On some level, the plan had also relied on the involvement of the public sector for funding, but this support failed to materialize. Additional challenges also hindered the realization of a bio-building material's supply chain, based on hemp grown on uncontaminated soils elsewhere in the valley. Organizing the process proved technologically and financially difficult. While some processing machinery existed, it was prohibitively expensive. This reflected a broader issue within contemporary Italian agriculture concerning the transformation of hemp, in terms of the available technologies and the financial costs of obtaining them. Moreover, producing enough biomass to supply even a semi-industrial processing chain proved unfeasible, given the limited agricultural production capacity of the alpine valley. These difficulties brought the discussions to a standstill.

Garden Three: Cultivating Hemp, Making a Living

Giovanni invited me to come up to his house before heading to his fields. Although I had known him for a few years, this was the first time I visited his hemp garden. Giovanni started cultivating hemp to reduce the proliferation of unwanted plants. Later, after joining the *Hemp&Hemp* association, he learned about the plant's food uses and began growing it for its seeds—a matter prized for its nutritional qualities. Today, he devotes about 2000 square meters of land—located not far from his home in the lower Susa Valley—to hemp cultivation.

Giovanni cultivates approximately 15,000 square meters (1.5 hectares) of scattered plots between the plain and the mountains. That makes him a relatively established pluriactive farmer in the context of local agriculture, combining income from different activities (Mignemi, 2020; Narotzky & Smith, 2006). Alongside hemp, Giovanni grows potatoes, vegetables, and wheat. He accesses the land through a variety of arrangements: sometimes by sharing part of the harvest with landowners, offering gifts at Christmas, or simply by keeping the fields in good order. For Giovanni, farming is part of a life trajectory marked by several changes: he has run a commercial business, organized tourist activities, and explored other ventures. He is skilled at navigating and experimenting with multiple activities, yet he tells me he turned down more profitable economic opportunities to preserve autonomy over what he does. He is seeking to slow down, to do less but do it better. This ethos informs his farming practices, such as his choice to grow potatoes without using insecticides or pesticides. Wage stability is not a priority either; rather, he values spending time with his family and the freedom to choose only the work contexts he enjoys.

Giovanni had invited me to join him for the hemp harvest. Before we left his house, he showed me that year's potato and wheat harvest. We then set off in his car towards the small hemp field, that is located in the urban outskirts, an area where urban development is less dense but still present. The town center is not far from the field, in a valley where industrial and urban areas alternate with small agricultural plots. When we arrived, I observed the plants: the stalks were slender and heavy with seeds. The stems had turned yellow, while the tops remained green. The whole field was bushy and dense, making it difficult to walk through. Giovanni had been cultivating the *Futura 75* hemp variety for several years—one of the certified strains available on the European market. For the harvest, he used a compact combine harvester. It was not easy for him to find a machine small enough to navigate the field. Moving and renting such equipment pose challenges for the local farmers given the elongated geography of the Susa Valley, where plots can be separated by 30 kilometers or more.

Giovanni started up the combine harvester and drove it slowly back and forth across the field. The machine sometimes struggled to cut through the plants' tough fiber. This happened a few times over the course of the two-hour harvest. As he continued, the seeds accumulated in the combine harvester's grain tank, damp and warm. The following day, Giovanni would clean them from the impurities and leave them to dry. The harvest yielded 19 kilos of food seeds. "Whole seeds sell for 28 euros a kilo, not bad", he told me. He was satisfied, even if the harvest was modest that year, and the earnings limited. After factoring in the cost of seeds, the fuel, and the packaging process, the hemp harvest will bring him only 200€ profit.[5]

In Giovanni's field, we see the only supply chain that has managed to produce and sell a final product—whole or hulled seeds, and hemp seed oil—for some years: the food supply chain. This particular supply chain benefits from relatively low processing costs. However, as Giovanni's experience suggests, it has struggled to produce enough seeds and generate a substantial income. Several factors have affected hemp seed yields across the plantations and during different seasons: excessive rainfall preventing germination, damage caused by birds on crops, poor harvests, and the limited cultivated area have all made it difficult to reach volumes that would allow food production to scale up consistently.

[5] As mentioned earlier, Giovanni is a pluriactive farmer and his incomes do not depend on hemp alone but are spread across multiple crops and activities. As such, hemp is just one element within a small-scale, diversified production.

Cultivating Niches, Imagining Alternatives

Anthropological work on alpine contexts has long examined the interplay between ecological, political-economic, and cultural dynamics (Cole & Wolf, 1974/1999; Netting, 1981; Rosenberg, 1988; see also Viazzo, 2018). These inquiries have been revitalized in recent years by new readings of the socio-ecological transformations in the Alpine region in the context of the Anthropocene (Krauß, 2024).[6] Building upon this scholarship, I followed the "hemp supply chains" project across the gardens I described above in order to study the peculiarity of the Susa valley: a territory marked by the ecological and economic differences between the lower and the upper valley. The lower valley bears the marks of several decades of industrialization and maintains close ties to the urban area of Turin. In contrast, the upper valley is characterized by a mountain tourism industry that has significantly shaped the local economy and ecology. Observing the hemp gardens and the contexts in which they are grown reveals different (and conflicting) political-economic projects existing within the same ecologies. Lucia and Enrico, for instance, promote an alternative model of tourism in the upper valley, in contrast to the dominant ski industry. Meanwhile, participants in the evening debate on hemp and phytoremediation envisioned the possibility of a bio-building materials supply chain that could remediate the environmental legacies of industrial development and pollution in the lower valley (Centemeri, 2021; Centemeri et al., 2022; Vaccaro et al., 2016).

The three gardens also reflect the presence of different social actors. Lucia and Enrico are part of a demographic trend reversal in the Italian Alps that began in the 2000s (Viazzo, 2023, p. 63), marked by the arrival of new inhabitants in the region (Corrado et al., 2014). In contrast, many of those attending the hemp and phytoremediation meeting are long-time residents of the lower valley. Some work locally while others commute regularly to Turin. Life in the Susa valley exemplifies how mountain, rural, and urban dimensions coexist and overlap daily, with residents routinely negotiating these spaces (see Dematteis, 2018).

Hemp also represents the continuity of a political-economic and political-ecological project, that of the "hemp supply chains", across

[6] More generally, the very tradition of anthropological "political economy" has contributed to the emergence of "political ecology", providing it with themes and research tools (Roberts, 2020).

distinct ecological niches. Although the small village in the upper valley is not well suited to hemp cultivation, for Lucia and Enrico, the act of growing it carries a symbolic weight. It allows them to maintain connections with others involved in the hemp association. In the lower valley, ecological niches are more favorable for hemp cultivation, yet prove insufficient to enable Giovanni to transform the hemp crop into a substantial contribution to his livelihood. Moreover, some (post)industrial areas are so ecologically degraded that they have been rendered forcibly unproductive. Inverting Cole and Wolf (1974/1999, p. 19), I suggest that, as hemp moves through the varied ecologies of the Susa valley, it reveals the ecological divergence between the hemp gardens and their ideological convergence. While this ideological convergence is not monolithic, "degrowth" approaches (Latouche, 2007) have nonetheless inspired at least some of the actors involved in the hemp project. The spread of cultural sensitivity to degrowth discourses is part of the legacy of the No TAV movement where the opposition to the new railway project has become intertwined with a broader critique of hegemonic discourses about economic development. Traces of this intellectual, political, and moral engagement are visible—albeit in different ways—in the three cases I have presented earlier: Lucia and Enrico's quasi withdrawal from the world of work (see Weeks, 2011); the efforts to make the (post)industrial territories of the lower valley productive in different terms, through an "ethical", small-scale, agriculture-based industry; and Giovanni's decision to slow the pace of his working life, subordinating economic activities to his personal and familial priorities, and avoiding the social pressures to accept any kind of job or to maximize production.

However, this ideological continuity is fraught with difficulties and failures. Growing hemp in the high mountains is constrained by the altitude, while cultivation in the lower valley encounters other limitations and setbacks. The limited availability of land in an alpine valley makes it difficult to supply a sustainable building materials industry. Even the food supply chain, which does not require a minimum production threshold, has struggled to stabilize. Over the years, adverse weather, environmental stressors, and poor harvests, have all hindered a consistent seed production. Processing has also posed problems. Although Giovanni was able to purchase a small combine harvester for his field, accessing affordable machinery has often been very complicated, as in the case of the bio-building materials supply chain.

Despite these difficulties and failures, in the eyes of *Hemp&Hemp* association members, hemp supply chains hold the potential to foster an alternative local economy, one rooted in local ecologies and distinct from the material and financial flows underpinning large-scale infrastructure projects, such as the TAV. Hemp promises to reconnect agriculture and (semi) industrial processing—potentially reconfiguring the relationship between production and the local environment, and re-signifying the region's industrial past (Cacchioni, 2021b). I suggest that the political-economic and political-ecological significance of these attempts can be understood in two ways. I interpret them as an effort to produce less alienated resources, drawing on Tsing's understanding of alienation, which expands on the Marxian notion:

> Through alienation, people and things become mobile assets; they can be removed from their life worlds in distance-defying transport to be exchanged with other assets from other life worlds, elsewhere. [...] Alienation obviates living-space entanglement. The dream of alienation inspires landscape modification in which only one stand-alone asset matters; everything else becomes weeds or waste. (Tsing, 2015, pp. 5–6)

Hemp could thus re-embed resource production and manufacturing within the entanglements of local living-space. Furthermore, the appeal to hemp as a catalyst for a new economy reflects a desire not to completely break with the industrial past. Therefore, what emerges is an attempt to "redirect" parts of the industrial heritage toward new economic and ecological arrangements (Bonnet et al., 2021).

In the final section, I will explore how these characteristics of the Susa Valley "hemp supply chains" project resonate more broadly with anthropological scholarship on economic alternatives, agriculture, and supply chains.

Conclusion: Highlighting the Connection Forms in the Making of Alternative Supply Chains

In this chapter, I have examined hemp cultivation across distinct ecological and economic niches in the Susa Valley. Hemp cultivation participates in a valley-wide project that seeks to establish local hemp processing chains. However, these efforts have so far had little success. The project has required participants to interpret and navigate diverse environments and to envision ways of engaging in productive connections centering on hemp. In this last section, I discuss three dimensions that are key to the creation of alternative supply chains: capital and technology; the relationship with the dominant economy; and ideology.

In the struggle to articulate an alternative to the TAV economy, capital and technology emerged as central dimensions in the analysis of collaborative supply chains projects, as connection projects between ecological and economic niches. These elements can be overlooked in the literature examining agriculture-based supply chains and alternatives to the dominant political economy (Tsing 2012, 2015). Yet they become especially salient when efforts are made to construct provisioning paths (Narotzky, 2005) which, while remaining local, seek to integrate agriculture with forms of industrial processing. In contrast, other forms of provisioning often work around these obstacles: Community-Supported Agriculture typically involves supply chains without much processing (Lamine, 2008); low-tech initiatives require minimal capital investment; and in the case of recovered enterprises, capital is often reclaimed from within the capitalist system itself (Berlioux, 2021).[7]

The difficulty of accessing technology and capital also points out the forms of connection between the "hemp supply chains" in the Susa valley and the hegemonic capitalist economy (Narotzky, 2012). While the economic and power imbalance in the valley between the promoters of the TAV and its opponents is materially overwhelming, it is nonetheless important to consider how the relationship with the dominant economy—and the multiple crises it brings to the valley—is perceived and experienced (Narotzky & Besnier, 2014). These crises include the economic fallout of deindustrialization, and the ecological damages caused by the construction of the TAV, as well as the pollution left by industry in the lower valley. The hemp supply chains emerge in *opposition* to previous productive activities, such as the ski industry and the polluting industries on the valley floor. At the same time, the hemp supply chains are also developed *at the margins* of the dominant productive activities: on residual land, in spaces not occupied by urban or industrial expansion. However, they do not represent a spatial escape from the dominant economy, like certain experiences seeking "alternative lifeworlds" in mountain

[7] This point also highlights a broader need for further reflection in economic anthropology on the role of the state and public policies in extending the access to capital and technology, especially through a diachronic and comparative perspective.

regions (Koop, 2020): they remain situated *within* the valley's ecology. In a sense, the hemp-related initiatives accept part of the industrial legacy, seeking to "repair" certain aspects of it and "redirect" some elements of the industrial heritage (Centemeri, 2021; Centemeri et al., 2022; Bonnet et al., 2021). The "hemp supply chains" case thus contributes to a literature that challenges a binary understanding of the relationship between the dominant and alternative economies (Narotzky, 2012; Siniscalchi, 2019). In particular, it highlights the importance of attending to plural and situated ways in which alternatives are both enacted and framed in connection to broader economic structures.

To conclude, I return to the ideological dimension of economic change (Müller, 2006). If the capitalist supply chains have shown the capacity to connect heterogeneous ideological and cultural niches (Tsing, 2009, 2012, 2015), what about alternative supply chains? In the Susa Valley, a shared background of critique of the TAV project and a broad socialization into degrowth discourses have played a role in enabling diverse actors to coordinate around the "hemp supply chains" initiatives. The strength of this cultural substratum lies in its character as a weak ideology (see Jarrige & Tordjman, 2023): flexible enough to accommodate multiple interpretations and enactments, as shown by the diversity of the three hemp gardens. Still, this ideological variability remains far more limited than the pronounced heterogeneity of the "freedom agendas" among mushroom pickers in the Oregon forests described by Tsing (Tsing, 2012, p. 41). This prompts us to consider how much more ideological coherence is required to establish alternative supply chains, in comparison to capitalist ones. And this raises also another question: what can (different elements of) ideology do? In the Susa Valley, hemp supply chains paradoxically reflect both economic marginality and ideological centrality. They are marginal in that they fail to scale up or achieve consistent productivity, remaining peripheral to the local economic life. Yet they are ideologically central—at least in their imagined potential—as they have originally been conceived not as fringe experiments but as instruments to reform and reshape the local economy. Compared to other ideological traditions within alternative economies, hemp supply chains offer a distinctive synthesis of systemic thinking and local initiative (DuPuis & Goodman, 2005; Schlosberg & Coles, 2016).

This study of "hemp supply chains" project in the Susa valley contributes to a broader understanding of the relationship between agriculture and extractivism, the central topic of this volume. In the Susa valley context, hemp farming configures an alternative to what may be perceived as an extractive project such as the TAV (Manes, 2018). Hemp farming is embedded within a local supply chains initiative that aims to sustain an economy capable of meeting human needs by sourcing plant resources from niche ecologies. In this sense, the "hemp supply chains" can be seen as a potential contribution to a post-extractive reconfiguration of material flows, one that seeks to integrate agriculture and small-scale processing.

Acknowledgments My reflection on supply chains has been enriched by some important moments of intellectual exchange. Together with Morgan Jenatton, I co-organized two events to reflect on "*filières*": a workshop during the *Biennale d'Ethnographie de l'EHESS*, held in Marseille on 26 and 27 September 2022; and a *journée d'étude* "Filières. Matérialités, imaginaires et connexions socio-écologiques dans les chaînes de production", which also took place at the EHESS Marseille campus, at the Centre Norbert Elias, on 30th November 2022 (https://centrenorbertelias.cnrs.fr/filieres_2022/). I would like to thank Morgan Jenatton for the rich exchange that made these events possible, and all those who contributed in various roles during both occasions. I am grateful to Gabriele Orlandi for our intense discussions about Alpine anthropology. I would also like to express my deepest gratitude to Letizia Bonanno for profoundly re-reading an earlier version of this text and to the editors, whose comments made possible a significant improvement of my chapter writing. Finally, I am very grateful towards my interlocutors in the Susa Valley, with whom I have had the privilege and the pleasure of sharing questions and moments of life.

References

Accornero, A. (2011). *Quando c'era la classe operaia. Storie di vita e di lotte al Cotonificio Valle Susa*. Società editrice il Mulino.

Aime, M. (2016). *Fuori dal tunnel. Viaggio antropologico nella val di Susa*. Meltemi.

Berlioux, F. (2021). La solidarietà operaia messa alla prova dall'individualizzazione del lavoro. Il caso della fabbrica Fralib (D. Cacchioni & E. Dinubila, Trans.). *Meridiana, 102*, 75–94.

Bermond, C. (2007). Sestrières 1930—1990. Una "villanova" contemporanea per gli sport invernali. In I. Lopane & E. Ritrovato (Eds.), *Tra vecchi e nuovi equilibri. Domanda e offerta di servizi in Italia in età moderna e contemporanea. Atti del quinto Convegno nazionale, Torino, 12-13 novembre 2004* (pp. 567–580). Cacucci Editore.

Berta, G. (2006). *L'Italia delle fabbriche. Ascesa e tramonto dell'industrialismo nel Novecento*. Società editrice il Mulino.

Bonnet, E., Landivar, D., & Monnin, A. (2021). *Héritage et fermeture. Une écologie du démantèlement*. Éditions divergences.

Cacchioni, D. (2021a). Collaborare, intraprendere, sperimentare: Coltivatori e associazioni rurali in Val di Susa. *Meridiana, 102*, 41–56. https://doi.org/10.23744/4253

Cacchioni, D. (2021b). Hemp and Industry in Italy: Between Pasts and Present. *AGER: Revista de Estudios sobre Despoblación y Desarrollo Rural, 32*, 93–115. https://doi.org/10.4422/ager.2021.09

Caruso, L. (2010). *Il territorio della politica. La nuova partecipazione di massa nei movimenti No Tav e No Dal Molin*. FrancoAngeli.

Centemeri, L. (2021). Riabitare. L'impegno ambientalista in un mondo rovinato. In G. Caudo & M. Pietropaoli (Eds.), *Riabitare il mondo*. Quodlibet.

Centemeri, L., Topçu, S., & Burgess, J. P. (2022). Introduction. Recovery, resilience and repairing: For a non-reductionist approach to the complexity of post-disaster situations. In L. Centemeri, S. Topçu, & J. P. Burgess (Eds.), *Rethinking post-disaster recovery. Socio-Anthropological perspectives on repairing environments* (pp. 1–21). Routledge.

Chiaramonte, X. (2019). *Governare il conflitto. La criminalizzazione del movimento No Tav*. Meltemi editore.

Cole, J. W., & Wolf, E. R. (1999). *The hidden frontier: Ecology and ethnicity in an Alpine Valley*. University of California Press. (Original work published 1974).

Corrado, F., Dematteis, G., & Di Gioia, A. (Eds.). (2014). *Nuovi montanari. Abitare le Alpi nel XXI secolo*. FrancoAngeli.

Dematteis, G. (2018). The Alpine Metropolitan-Mountain Faced with Global Challenges. Reflections on the Case of Turin (Translated by the author and revised by Luigi Genta, Trans.). *Journal of Alpine Research | Revue de Géographie Alpine, 106–2*. https://doi.org/10.4000/rga.4402

Dubuisson-Quellier, S. (2018). *La consommation engagée*. Presses de Sciences Po.

DuPuis, E. M., & Goodman, D. (2005). Should we go "home" to eat?: Toward a reflexive politics of localism. *Journal of Rural Studies, 21*(3), 359–371. https://doi.org/10.1016/j.jrurstud.2005.05.011

Grasseni, C. (2014). Seeds of Trust. Italy's Gruppi di Acquisto Solidale (Solidarity Purchase Groups). *Journal of Political Ecology, 21*(1), 178–192. https://doi.org/10.2458/v21i1.21131

Jarrige, F., & Tordjman, H. (2023). Actualités de la décroissance. In F. Jarrige & H. Tordjman (Eds.), *Décroissances* (pp. 5–20). Le passager clandestin.

Koop, K. (2020). Escaping from capitalism: The enactment of alternative life-worlds in France's mountain regions. In S. M. Hall, H. Pimlott-Wilson, & J. Horton (Eds.), *Austerity across Europe* (pp. 125–140). Routledge.

Krauß, W. (2024). Epilogue. Alpine anthropology in the anthropocene. In A. Schneider & E. Tauber (Eds.), *New Horizons for the Alps. Ethnographies, reshaping challenges, and emerging more-than-alpine relations* (pp. 285–300). Bozen-Bolzano University Press.

Lamine, C. (with Perrot, N.). (2008). *LES AMAP: un nouveau pacte entre producteurs et consommateurs ?* Éditions Yves Michel.

Laszczkowski, M. (2019). Encountering 'micro-dust': Material entities, affect, and politics in infrastructural conflict in Italy. *Ethnos, 85*(5), 938–956. https://doi.org/10.1080/00141844.2019.1640758

Laszczkowski, M. (2025). The big, the small, and the ugly: The politics of scale-making in a contested railway project in Italy. *Ethnos, 90*(2), 247–264. https://doi.org/10.1080/00141844.2023.2240541

Latouche, S. (2007). *La scommessa della decrescita*. Feltrinelli.

Leonardi, E. (2013). Foucault in the Susa Valley: The No TAV Movement and Struggles for Subjectification. *Capitalism Nature Socialism, 24*(2), 27–40.

Manes, L. (2018, August 19). Zibechi: «Dagli zapatisti ai No Tav, lotta globale all'estrattivismo». *il manifesto*. https://ilmanifesto.it/zibechi-dagli-zapatisti-ai-no-tav-lotta-globale-allestrattivismo

Martini, G. (2016, June 3). *Sarà la cannabis a salvare Taranto dal disastro ambientale dell'Ilva—La Stampa*. https://www.lastampa.it/cronaca/2016/06/03/news/sara-la-cannabis-a-salvare-taranto-dal-disastro-ambientale-dell-ilva-1.34981334/

Mignemi, N. (2020). Pluriattività rurale e lavoro agricolo: Note introduttive per un cantiere (ancora) aperto. In N. Mignemi, C. Lorenzini, & L. Mocarelli (Eds.), *Pluriattività rurale e lavoro agricolo in età contemporanea (secoli XIX-XX)* (pp. VII–XXII). New Digital Frontiers.

Müller, B. (2006). *Disenchantment with Market Economics. East Germans and Western Capitalism*. Berghahn Books.

Narotzky, S. (2005). Provisioning. In J. G. Carrier (Ed.), *A handbook of economic anthropology* (pp. 78–93). Edward Elgar Publishing.

Narotzky, S. (2012). Alternatives to expanded accumulation and the anthropological imagination. turning necessity into a challenge to capitalism? In P. G. Barber, B. Leach, & W. Lem (Eds.), *Confronting capital: Critique and engagement in anthropology* (pp. 239–252). Routledge.

Narotzky, S. (Ed.). (2020). *Grassroots economies: Living with austerity in Southern Europe*. Pluto Press. https://doi.org/10.2307/j.ctv17ppccg

Narotzky, S., & Besnier, N. (2014). Crisis, value, and hope: Rethinking the economy: An Introduction to supplement 9. *Current Anthropology, 55*(S9), S4–S16. https://doi.org/10.1086/676327

Narotzky, S., & Smith, G. (2006). *Immediate struggles. People, power, and place in rural Spain*. University of California Press.

Netting, R. M. C. (1981). *Balancing on an Alp. Ecological change and continuity in a Swiss mountain community*. Cambridge University Press.

Oliva, G. (2020). Dal ventennio fascista alla democrazia: Sindacato, partiti e movimenti politici. In P. Del Vecchio & D. Vota (Eds.), *Storia delle valli di Susa dall'Unità d'Italia agli anni Settanta del Novecento* (pp. 17–42). Edizioni del Graffio.

Pepino, L., & Revelli, M. (2012). *Non solo un treno... La democrazia alla prova della Val Susa*. Edizioni Gruppo Abele.

Poni, C., & Fronzoni, S. (2005a). Premessa. In C. Poni & S. Fronzoni (Eds.), *Una fibra versatile. La canapa in Italia dal Medioevo al Novecento* (pp. VII–X). CLUEB.

Poni, C., & Fronzoni, S. (Eds.). (2005b). *Una fibra versatile. La canapa in Italia dal Medioevo al Novecento*. CLUEB.

Roberts, J. (2020). Political ecology. *The Open Encyclopedia of Anthropology*. https://www.anthroencyclopedia.com/entry/political-ecology

Rosenberg, H. G. (1988). *A negotiated world: Three centuries of change in a French alpine community*. University of Toronto Press.

Sacco, S. (2002). *ASSA. Acciaieria di Susa*. Edizioni del Graffio.

Sacco, S. (2020). Le condizioni economiche del Circondario di Susa a cavallo tra Ottocento e Novecento. In P. Del Vecchio & D. Vota (Eds.), *Storia delle valli di Susa dall'Unità d'Italia agli anni Settanta del Novecento* (pp. 43–84). Edizioni del Graffio.

Schlosberg, D., & Coles, R. (2016). The new environmentalism of everyday life: Sustainability, material flows and movements. *Contemporary Political Theory, 15*, 160–181. https://doi.org/10.1057/cpt.2015.34

Siniscalchi, V. (2019). Mobilisation, activism and economic alternatives. In J. G. Carrier (Ed.), *A research agenda for economic anthropology* (pp. 105–118). Edward Elgar Publishing.

Tsing, A. L. (2009). Supply chains and the human condition. *Rethinking Marxism, 21*(2), 148–176. https://doi.org/10.1080/08935690902743088

Tsing, A. L. (2012). Empire's salvage heart: Why diversity matters in the global political economy. *Focaal, 2012*(64), 36–50. https://doi.org/10.3167/fcl.2012.640104

Tsing, A. L. (2015). *The mushroom at the end of the world: On the possibility of life in capitalist ruins*. Princeton University Press.

Vaccaro, I., Harper, K., & Murray, S. (Eds.). (2016). *The anthropology of postindustrialism: ethnographies of disconnection*. Routledge.

Viazzo, P. P. (2018). Tre fasi nella storia dell'antropologia alpina. In G. Ligi, G. Pedrini, & F. Tamisari (Eds.), *Un accademico impaziente. Studi in onore di Glauco Sanga* (pp. 261–273). Edizioni dell'Orso.

Viazzo, P. P. (2023). Dagli «immigrati nei villaggi degli emigranti» ai «nuovi montanari». Mobilità di sostituzione nelle Alpi prima e dopo l'età dello spopolamento. *Popolazione e storia, 24*(2), 57–74. https://doi.org/10.4424/ps2023-9

Weeks, K. (2011). *The problem with work. Feminism, Marxism, antiwork politics, and postwork imaginaries.* Duke University Press.

Who is Served by Agricultural Cooperatives? Power Relations in the Lleida Fruit Cluster (Catalonia, Spain)

Jordi Gascón

Introduction

The Spanish economy underwent structural change in the late 1970s. Spain was preparing to join the European Economic Community and had to adapt to European standards. In the agricultural sector, this meant accelerating the processes of technification and organisation that had begun in the 1960s, when the period of autarchy under the Franco regime came to an end (Barciela, 2017). Agricultural cooperatives played an important role in this paradigm shift. Like most Western cooperatives, Spanish cooperatives focused on services and sales. Their main purpose was to centralise sales of their members' products and purchase supplies collectively, such as seeds, agrochemicals, machinery and fuel. These days, they also participate in the coordination of agricultural extension programmes aimed at training farmers in emerging market demands. In this

J. Gascón (✉)
Department of Anthropology, University of Barcelona, Barcelona, Spain
e-mail: jordigascon@ub.edu

S. Narotzky et al. (eds.), *Agricultural Extractivism in the Mediterranean Region*,
https://doi.org/10.1007/978-3-032-05599-6_13

new, more complex and competitive context, cooperatives have become more concentrated and more corporate (Juliá et al., 2013). This chapter considers whether this shift in the agricultural sector allowed cooperatives to strengthen the role of producers in the agrifood system or whether it made them into a tool for controlling and subordinating farmers.

The academic literature on modern agricultural cooperatives in Spain has largely presented them as an instrument that helps connect producers to markets and makes them more competitive (Ruiz et al., 2006; Campos-Climent & Chaves-Ávila, 2012). Similarly, several studies have concluded that cooperatives give farmers greater negotiating power with large distributors (Monzón, 2003; Valentinov, 2007). The literature also identifies a series of problems affecting cooperatives, such as their inability to set prices, their reliance on sales channels that periodically go into crisis, the disaffection of their members and the impossibility of sharing profits. These difficulties are understood as "challenges" that may be overcome through efficient, flexible management to meet market demands (Ruiz et al., 2013; Meliá & Juliá, 2008).

Few studies have approached these supposed "challenges" as structural problems caused by an unequal agrifood chain that is controlled by large distributors and the agrochemical sector (McMichael, 2013). Of the few scholars to have adopted such a perspective, the work of Martín et al. (2021) on cooperatives in the Canary Islands is particularly relevant. The authors conclude that cooperatives have been co-opted by large distributors and landowners, which have subordinated cooperative members (farmers) to their interests.

The research presented in this chapter takes a similar approach. During our fieldwork in the province of Lleida (Catalonia, Spain), we observed that cooperatives are receiving constant complaints from their members for failing to fulfil their founding objectives. We consider whether this recurring criticism is linked to power differences between producers and large distributors (supermarket chains, trade associations with a single purchasing centre and exporters).

Lleida is the largest apple and pear-growing area in the Spanish State, as well as one of the largest peach, nectarine and flat peach areas. Fruit production is concentrated around the city of Lleida, the provincial capital, in the counties of Segrià, Pla d'Urgell, Urgell, Noguera and Garrigues (Fig. 1). In 2022, it occupied an area of 27,178 hectares, almost all of which is irrigated (DARP, 2024).

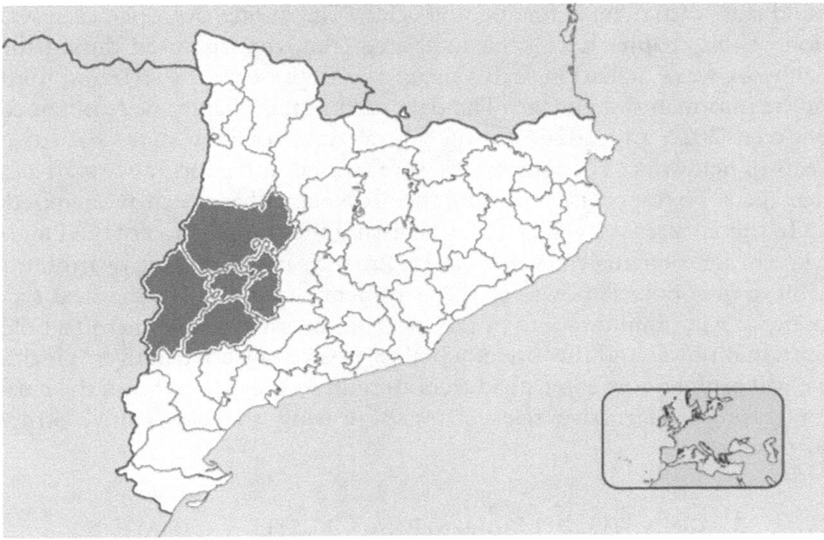

Fig. 1 Largest tree fruit production areas in Lleida. (Source: author)

The three main actors in the Lleida fruit cluster are farmers, collection centres (companies and cooperatives) and large distributors.[1] As is now common in the global agrifood system (Reardon et al., 2003; McMichael, 2009), large distributors have become the hegemonic actors in the Lleida cluster.

An ethnographic methodology was used for our research, with the aim of uncovering and understanding social behaviour using qualitative research techniques: participant observation, semi-structured interviews and focus groups. This methodology allows individual narratives to be analysed in relation to their social, historical and ideological context. In-depth interviews were conducted to (re)construct the experiences of the informants in detail (Robles, 2011) and the researchers sought to

[1] Family farmers work the land in Lleida, with support from hired labour at specific points in the agricultural cycle. These seasonal workers experience working conditions characterised by exploitation, marginalisation and racialisation. The relationships between farmers and seasonal workers are beyond the scope of our analysis, but they have been studied by a number of researchers, including Gonzálves and López Trigal (1999), Casal (2019), Mata (2018) and González et al. (2021).

build trust with them before commencing. The quotes that open each section of this chapter have been translated (the language used during the fieldwork was Catalan) in such a manner as to preserve the style and spirit of the informants' language. The data used in this chapter were obtained between 2016 and 2024, during several stays totalling more than two years of fieldwork. The informants were farmers and cooperative staff and managers. To respect their privacy, their real names have been anonymised.

In this chapter, we will begin by analysing the composition of the Lleida fruit cluster. Our analysis will focus on the ways in which large distributors achieve oligopolistic power to the detriment of farmers. In the next section, we will examine some of the mechanisms that enable them to hold onto that power and the instrumental role played by cooperatives. Finally, we will explore how some producers attempt to break away from the cluster through alternative sales channels, leaving the cooperative structure behind.

Productive Specialisation, Cluster Formation and Power Relations

> The cooperative world… It's a very good idea, the essence is very good. I'm a committed cooperativist. I've advocated for it constantly. And now I'm tired […] It's a model that farmers should defend, but you end up realising that it doesn't work, despite 90% of farmers having [their produce] in the cooperative […] It's like any other warehouse, but on top of that it gives you poor economic outcomes because it's so badly managed. (Francesc, fruit farmer)

Until the late 1950s, agriculture in central Lleida was diverse. Fruit and vegetables were grown on irrigated land, while cereal crops, fodder, olive trees and cattle farming were predominant on non-irrigated land (Bretón Solo de Zaldívar, 2000). Not long after, tree fruit farmers began to specialise in pear, apple and peach farming. This change was due to several factors. On the one hand, a growing domestic market that demanded products that were once unaffordable for the majority of the population, like fruit. On the other, access to the European market after the Franco dictatorship abandoned its policy of economic autarchy and Western countries began to accept Spain in the international order. At a time when the purchasing power of the European population was on the rise, fruit from Lleida was very popular in Europe due to the favourable exchange rate of the Spanish currency and low production costs. In short, access to the European and domestic market decoupled producers from local markets and prompted them to specialise in fruit farming (García Manrique, 1971; Bretón Solo de Zaldívar, 2000).

In 1986, Spain joined the European Economic Community. Its accession marked the end of customs protection and the acceptance of European agricultural directives, making it necessary to restructure the agricultural sector (Clar, 2017). The agricultural landscape in Lleida did not change: it continued to centre around fruit. In fact, Europe encouraged productive specialisation by region (Soronellas & Casal, 2014), but competition with other fruit-producing countries (e.g. Italy and France) obliged farmers to lower production costs and new demands regarding stocks and quality led to the technification of production, storage and distribution (Langreo, 2012). The agricultural sector in Lleida adapted to these demands by making structural changes, such as irrigating all land, establishing a refrigeration network (Ruiz et al., 2003) and merging companies.

This business concentration occurred at two levels: in land ownership, with small farms unable to cover modernisation costs, and in sales. Small and medium-sized intermediaries vanished, making way for larger distribution structures: exporters, supermarket chains and trade associations with a single purchasing centre. In addition, the collection centres set up to link production and distribution were reinforced. They took a variety of forms: private wholesalers, agricultural transformation societies (*sociedades agrarias de transformación* SAT)), which were a variant of private wholesalers,[2] and cooperatives. Farmers, collection centres and large distributors emerged as the actors involved in the new fruit cluster in Lleida (Gascón et al., 2021) (Fig. 2.).

The fruit farming area around Lleida is home to the largest concentration of agricultural cooperatives in Catalonia (Vives-Xiol, 2019). Their aim is to unite producers to allow them to optimise production and build stronger links to the market. The strategies used to achieve this are joint marketing and promotion, collective acquisition of supplies and machinery, improved access to information and knowledge and collective bargaining vis-à-vis the public sector. The new cooperatives that were founded

[2] Agricultural transformation societies (*sociedades agrarias de transformación* (SAT)) are civil societies with an economic and social purpose, which centre around agricultural production, processing and sales. An SAT must have a minimum of three members. In Lleida, most SATs are family owned.

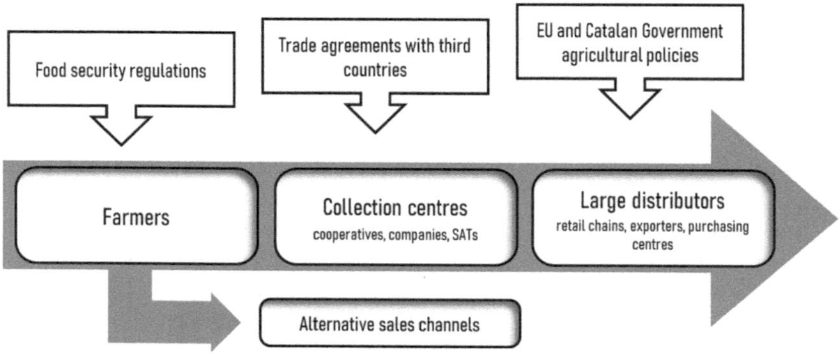

Fig. 2 The fruit cluster in Lleida since the 1980s. (Source: author)

from the late 1970s onwards presented a dual narrative. On the one hand, they aimed for efficiency: access to supplies, cutting-edge technology and training for farmers would help connect them to increasingly competitive globalised markets. On the other, they would be more influential: grouping farmers in cooperatives would boost their capacity to negotiate with other actors in the agrifood chain and public institutions.

However, these expectations were gradually thwarted. Large distributors became the dominant actors in the new cluster, rather like they did around the world (McMichael, 2009; Isakson, 2014). The factors explaining the situation in Lleida are present globally. Consumer habits changed and people passed over small grocery stores in favour of supermarket and greengrocer chains. Participation in an increasingly global market distanced farmers from the market, increasing the hegemony of large distributors (Farré & Sala, 2014). Finally, large distributors took over such a high percentage of fruit sales in both the domestic and international markets that farmers faced a bottleneck whereby most of them could only access markets through these distributors. This gave rise to a relationship based on dependency that allowed large distributors to impose prices and conditions on producers.

The functioning of the cooperatives was shaped by these unequal power relations. Thanks to their hegemonic power, the large distributors had no need to establish their own structures to collect and standardise the produce grown by the fruit farmers. Instead, they outsourced these activities to cooperatives, which were responsible for gathering the farmers' produce, managing stocks and monitoring quality. Over time, the cooperatives were obliged to merge to meet the logistical needs of a small number of intermediaries that covered vast, complex markets. This led to the emergence of second-degree cooperatives, whose management was even further removed from producers.

Our hypothesis is that, in this context, cooperatives became tools allowing large distributors to control and manage fruit production. These distributors took advantage of their hegemony over the cluster to impose conditions on farmers. Rather than doing so directly, they harnessed cooperatives to act as their supervisory agents.

Mechanisms for the Control and Subordination of Farmers

> Large retailers, their purchasing power allows them to do whatever they want. Anyone who says different is lying. I'm telling you the truth, they do whatever they want. "I want pears up to 15 July". "How many?" "As many as I can sell". (Joan, cooperative manager)

Collection centres and, more specifically, agricultural cooperatives, perform a twofold function for large distributors: they allow them to externalise operating costs to the producer and to monopolise as much of the profit from agricultural production as possible. In the rest of this section, we will present three mechanisms that contribute to these ends: quality standards for produce, constant upgrading of farms through varietal change and pricing.

Quality and Appearance

When assessing the quality of fruit, conventional sales channels prioritise its appearance over its nutritional value or organoleptic properties. Visual irregularities and size differences are unacceptable, despite them having no effect on nutritional quality. Fruit is also expected to be at the same stage of ripeness. Farmers and cooperative managers are aware that their produce must be homogeneous in terms of size, texture, colour, shape and ripeness in order for it to be accepted by the market, that is by large distributors.

These requirements are imposed by large distributors, which seek to standardise the produce entering the agrifood system (Gorenstein, 1998;

Prieto et al., 2008). They are not an indulgence; rather, they are the outcome of the logistical needs of distributors. To maintain good stock control or facilitate exports, all fruit must ripen at the same rate after being harvested and this is only possible if the products are a similar size or at the same stage of ripeness. Another logistical need has to do with the sales system employed. At supermarkets and greengrocer chains, the task of selecting products and preparing bags has been outsourced to consumers. If the fruit varies widely in appearance, consumers will select the most appealing fruit and the rest will be left and eventually rot (Gascón et al., 2022; Gascón et al., 2023).

Cooperative managers are aware that the logistical system for the cluster has been shaped by the hegemony of large distributors. This power allows distributors to shift part of their operating costs to cooperatives and farmers. They establish which produce they will and will not purchase according to their quality standards, obliging cooperatives to oversee their requirements.

These quality standards are not always the same. If the harvest is poor, smaller, more irregular fruit will be accepted. Similarly, at the start and end of the season, when there is little produce on the market, quality controls are relaxed. By contrast, during years of abundant harvests and at the peak of the harvest, quality standards are stricter. Therefore, there is no fixed rule establishing what produce is accepted. Producers explain that produce is sometimes harvested according to a specific size, but that they cannot be certain that the cooperative will not stipulate a different size at a later date upon the instructions of the large distributor.

Ultimately, three quality filters are applied to the produce before it reaches the supermarket or shipping container for export. The first is carried out on the farm by the farmer. Before starting to harvest, the farmer gives the pickers instructions regarding the quality of the fruit based on the requirements established by the cooperative, which has received them in turn from the large distributor. During the harvest, rotten fruit and fruit that has been attacked by insects and fungi is removed or left on the tree, along with fruit that is completely edible but does not meet the requirements for appearance and size. Fruit that would not survive the transport and storage process because it is too ripe is also rejected. Farmers may sell part of this discarded produce on secondary markets (cremogenates, farmers' markets). However, these markets are not powerful enough to absorb all their produce and a large amount of fruit is rejected on the farms themselves.

The second filter is applied by the cooperative. The cooperatives remove fruit that fails to meet the quality requirements set by the large distributors. Cooperative members frequently complain that cooperative managers fine farmers who bring produce that is smaller than the required size. Sometimes, they may have abided by the rules set by the cooperative, but the large distributor has made the requirements stricter at the last minute, for example by requiring a larger minimum size. To ensure that produce that does not meet the requirements is rejected, the fruit undergoes two quality controls at the cooperative. Monitoring of farmers by cooperatives begins on the farm, but is particularly exhaustive at the warehouse.

The third filter is applied by the large distributors to the merchandise that they receive from the cooperatives. Given the unequal contractual relationship between them and the cooperatives' fear of losing their client, the quality of the produce that reaches the large distributor tends to meet all the requirements. Cooperatives prefer to put pressure on farmers rather than negotiating with the large distributors upon which they depend.

Varietal Change

Varietal change has become a never-ending process. Few farmers can remember the varieties that were grown when fruit farming in Lleida became specialised, such as the Rome Beauty apple or the Blanca de Aranjuez pear. The peach varieties that were dominant in the 1970s, including Sundanell and Torres, have been replaced by Catherine and Baby Gold. The Limonera pear, which many producers consider a high-quality product and was predominant until recent decades, has made way for the Conference or Williams pear. This dynamism not only affects the varieties grown, but also the type of fruit. Between 2005 and 2017, 8,200 hectares of apple and pear trees were uprooted and replaced with peach and nectarine trees (Segre.com, 2018).

Farmers and cooperative managers identify two reasons for this varietal dynamism. On the one hand, the need to adapt to the market: to changing consumer tastes, but especially to the logistical needs of large-scale distribution. On the other hand, the need to reduce production costs, particularly labour costs.

In relation to the first group of causes farmers and cooperative managers describe two trends that have emerged in recent decades: rapid changes in the varieties of stone fruit grown (peach, nectarine, flat peach) and a reduction in the area allocated to pear and apple trees. They attribute these trends to changes in consumer tastes: new varieties emerge that are tastier, more visually appealing or easier to eat. For example, many claim that the success of the Paraguayan is due to its flattened shape, which allows for a cleaner bite. However, several studies attribute the changes in consumer taste to the ability of large distributors to impose trends (Montagut & Dogliotti, 2006; Ploeg, 2008). The aim would be to increase sales to increase sales (e.g. by expanding the fruit season) and solve technical and logistical problems. Farmers mention the case of peach trees as a paradigmatic example of this situation. Until the 1980s, only yellow peaches were grown. However, this variety caused technical issues for supermarkets and exporters: its skin is sensitive to handling and transportation. Currently, red peaches are predominant as they are much more robust. Farmers had to change varieties to meet the logistical needs of the large distributors, although they consider the organoleptic quality of yellow peaches to be better.

Yellow peaches are just one example. Farmers are obliged to continuously adopt new technological packages, which include varieties developed to meet the logistical and technical needs of large distributors. Cooperatives oversee this constant restructuring of farms through their agricultural extension work and efforts to enable access to loans.

The useful life of an apple tree is half a century, but they last no longer than 15 years in Lleida (Millán, 2009). The ongoing changes to the varieties and types of fruit grown mean that the useful life of the trees is not fully taken advantage of. Generally, varietal changes are carried out through grafts. In this case, trees may not produce fruit for one or two years and the cost is relatively low. On other occasions, the trees must be uprooted. The cost of purchasing seedlings is high: in 2019, it cost between 4,500 and 5,500 euros to purchase enough apple seedlings to cover a hectare, in addition to operating costs and lost production while the trees mature. Apple trees take five or six years to reach productive maturity. The costs all fall to the producer. Sometimes, they may receive public funds managed by the cooperative, which also imposes or facilitates the varietal changes demanded by the large distributors. For large distributors, the cost of varietal changes is zero despite them imposing and reaping the greatest benefits from such changes.

We said that there were two reasons for the varietal change. One, already analysed, is related to the demands of large-scale distribution. The other is the grower's need to reduce labour costs.

It is not only varieties that are emerging to solve the logistical problems of the large distributors. There are also varieties that aim to reduce human labour. These are varieties that require less thinning, reducing the investment in pruning, or that have a more grouped ripening, which reduces the number of harvesting passes (Miarnau Domènech, 2009). Again, the case of peaches is exemplary. Since the 1990s, trees with less volume and height have been sought so that the fruit can be harvested without a ladder. This reduces the number of hours worked (Montserrat et al., 2004).

We could ask ourselves whether the savings in labour force compensate for the cost of varietal change (time without production and price of the new technological package). But it is not worth it: the control of the cluster by large-scale distributors also allows them to set the purchase price of the production. The margin left to the farmer is minimal. Considering that it also does not control the price of technology packages or inputs, producers claim that the only variable over which they have control is labour. This allows them to assert that reducing the cost of labour to the minimum possible is not an option; it is a necessity in order to balance the farm's economic equilibrium.

The price of production is thus the third element in which the producer's subordination to large-scale distribution, mediated by the cooperative, materialises.

Pricing

Pricing is the third mechanism by which producers are subordinated to large distributors through cooperatives.

As noted, the hegemony of large distributors owes to their control over the market: most consumers purchase produce at supermarket chains, while exports, which are more and more logistically complex, can only be carried out by large export companies. The situation is made more difficult still by the rise of fruit farming in the European Union due to its expansion to Eastern European countries with a strong agricultural sector and by the signing of trade agreements with third countries allowing them to import fruit into the European market at competitive prices.

This strong competition allows large distributors to set prices. Prices are established according to the abundance or scarcity of certain produce on the European market. It is not unusual to see large distributors pay less than it costs farmers to produce the fruit (Miarnau Doménech, 2006; Iglesias & Casals, 2011).

In this manner, large distributors absorb most of the profit. In the case of fruit farming in Lleida, the difference between the farmgate price and the retail price was around 450% in the early 2020s (Table 1). It is important to remember that this is fresh fruit that requires no post-harvest processing: the farmgate price must cover the production costs and the difference between the farmgate and retail price only needs to cover transport and marketing costs.

Farmers only find out the price of their products when they take them to the cooperatives and are informed by the cooperative managers, but the decision on pricing is made by the large distributors. If the cooperatives express discontent, the large distributors will look for a different supplier. They even have the option to import produce. By contrast, the concentration of the distribution sector makes it very difficult for cooperatives to find other clients; if they do not bend to the demands and prices set by large distributors, they may find themselves cut off from the market.

Cooperative activity should generate profits for members, but this ceased to be the case at the end of the twentieth century due to pricing pressure by large distributors, which prevents cooperatives from making a healthy profit (Sabaté et al., 2000). Moreover, cooperatives must use their slender profits to modernise their infrastructure to meet market demands.

Table 1 Index of prices at origin and destination (IPOD) for the main fruits grown in Lleida and sold fresh (2023)[a]

Product	*Month*	*Price at origin (euros/kg)*	*Price at destination (euros/kg)*	*IPOD*
Apple	October	0.62	2.50	4.03
Pear	September	0.73	2.55	3.49
Peach	August	0.68	3.01	4.43
Nectarine	August	0.63	3.26	5.17

Source: Table compiled by the author based on COAG data (COAG, 2024)

[a]The Index of Prices at Origin and Destination (IPOD) is the number of times that the farmgate price is multiplied before reaching the consumer. In the table, the month in the middle of the season for each fruit is analysed

Leaving the Cluster, Fleeing the Cooperative

> We received messages by phone from the cooperative. When we received them, it was like "Godammit!". Because of the quality controls, you know? Now we're part of the organic circuit where everything goes… even scarred produce, small produce, it all gets sold! (Blai, organic producer)

In the chapter so far, we have analysed the thickest arrow in Fig. 1, which illustrates the fruit cluster in Lleida, but there is also a thinner arrow that represents alternative sales channels that have emerged in the last 20 years. Although they are growing, they constitute only a small proportion of total fruit production, accounting for less than 3% in 2019 (Observatori de l'agricultura i l'alimentació ecològiques, 2020). Nevertheless, they are interesting because they attract farmers who seek the autonomy that they have lost in the conventional cluster to ensure that their own farms are viable.

This alternative market revolves around organic production,[3] which is a strategy that allows a growing number of once conventional farmers to leave the cluster behind. However, it is not the mere fact of producing organic food that enables autonomy. Rather, organic production is a vehicle for accessing certain sales channels, which offer farmers autonomy (Gascón et al., 2021).

Albert is an organic farmer who manages 7.5 hectares of land: he grows fruit trees on 7 hectares and table grapes on half a hectare. He sells part of his produce to consumer cooperatives and part to networks of producers that sell produce at farmers' markets or deliver it to people's homes as part of a basket system. Andreu has 27 hectares of organically certified fruit trees, most of which are pear or apple. He also has five hectares of cereal crops, which he farms according to organic principles but are not officially certified. He sells all his produce through a collection centre specialising in the distribution of organic products.

The cases of Albert and Andreu illustrate the sales channels available outside conventional clusters. One option is to sell direct to end customers. Albert does so through consumer cooperatives, which are groups of consumers that establish commercial agreements with farmers directly. Another option is to sell produce at farmers' markets. Direct sales oblige farmers to spend time on marketing, but they allow them to take home all

[3] Farmers may or may not be officially certified, provided that they have generated sufficient confidence among consumers that their processes are organic.

the profits earned. In Lleida, there are fruit and vegetable farms covering less than two hectares that are viable because they sell all their produce directly to end customers. By contrast, conventional fruit farms often struggle to remain viable with 40 or 50 hectares of land.

Andreu sells his produce through an intermediary: an organic collection centre. For now, the relationships between organic farmers and distributors are more equal than those between conventional farmers and distributors. This is due to the fact that they do not have oligopolistic control over the market for several reasons: there are alternative sales channels for organic produce; supply and demand are less mismatched in the organic market than in the conventional market, where there are major issues with overproduction; and equal relationships with producers are part of the business policy and strategy. For various reasons, organic distributors work with a much lower Index of Prices at Origin and Destination (IPOD) than conventional distributors.

The viability of farms that seek out alternative sales channels does not owe solely to their greater share of profits or to the higher prices of organic produce (Aschemann-Witzel & Zielke, 2017). It is also due to the quality standards applied by the market. The organic market takes a different approach to quality than the conventional market, allowing more produce to enter the circuit. In the organic sector, the quality of produce is not linked to its visual appearance and the logistical structure of the distribution process does not require such a high degree of homogeneity, making a greater proportion of the produce marketable. Organic farms tend to have lower yields as they do not use chemicals to control pests, but this is compensated by the fact that the organic circuit accepts produce that would be rejected by the conventional market.

Therefore, these alternative sales channels offer two main benefits for farmers: they allow them to obtain larger profits and they do not put pressure on them to adopt a certain model of farming. This is because the relationships between the actors in this alternative cluster are more equal. They are likely to remain this way as long as a balance is maintained between production (supply) and consumption (demand) and there is not a single sales channel that affords intermediaries oligopolistic power. As the organic produce niche grows, large distributors are showing more interest in controlling it. Some certified organic produce is already being co-opted by these distributors through cooperatives, bringing it into the conventional cluster.

Finally, many of the farmers who have implemented this paradigm shift are reluctant to join any kind of stable association. Instead, they group together to achieve specific objectives. These alliances may become long-term initiatives, but they remain circumstantial, informal and aimed at achieving specific goals. Their members tend to be producers who have experienced working with cooperatives and have grown with them but eventually opted to leave them behind when they began to view them as structures that are instrumentalised to serve the interests of large distributors.

Conclusion

Spanish agriculture began to open up to the global market at the end of the 1950s, but this process grew more rapid with Spain's accession to the European Union. This triggered a restructuring of the agrifood system, which had a series of consequences. Two of those consequences are of particular interest to us here.

On the one hand, the distribution sector became concentrated and distributors gained oligopolistic power over the agrifood chain. In an increasingly complex sales structure, large distributors became a bottleneck between farmers and consumers, giving them unprecedented power (Montagut & Dogliotti, 2006; McMichael, 2009).

On the other, farms underwent major changes. The global market was incompatible with productive diversification and its logistical needs prompted it to transform land to make it more specialised (Abson, 2019). In the case of fruit farming in Lleida, this process is ongoing. Conventional farms must accept the cost of continuously incorporating new varieties and species to suit the interests of large distributors, whose hegemonic role allows them to burden farmers with these costs and to co-opt most of the profits.

However, the large distributors in the fruit cluster in Lleida (which include exporters, supermarkets and trade associations with a single purchasing centre) do not exercise this control directly. Remaining loyal to their policy of externalising processes and costs, they entrust this task to a key actor: agricultural cooperatives:

The new generation of cooperatives that appeared in the 1980s in the wake of the fruit farming boom were intended to enable producers to better access the emerging global market, improve their capacity to negotiate with distributors and strengthen their influence over public policy. The reality was quite the opposite. The direction of influence was reversed due to the hegemony of large distributors. In the end, cooperatives became a vehicle by which large distributors convey their requirements and interests to farmers. Cooperatives are obliged to accept this role to avoid being cut off from the market.

The organic niche offered an opportunity to leave the conventional fruit cluster behind and join alternative markets. It is a small market that absorbs only a limited proportion of the producers in Lleida, but that provides different sales channels and a more equal relationship between producer and intermediary. It even facilitates direct sales to end customers. This allows farmers to maintain more autonomy over the management of their farms and to obtain higher profits.

As Campbell (2009) explains, food regimes are the outcome of dialectical relationships between different logics, strategies and interests: although farmers play a subordinate role in the agrifood circuit, they do have options. Indeed, the search for alternative markets is not unique to Lleida. The quest for autonomy by abandoning the hegemonic agrifood system is a strategy employed by farmers around the world (Ploeg, 2008; Akram-Lodhi & Kay, 2010).

In Lleida, farmers who have left the conventional cluster behind establish agreements with one another and develop joint projects. It is relevant to note, however, that these are circumstantial relationships linked to specific goals. They may last over time, but they are contextually rooted. They reject the use of any kind of cooperative structure to coordinate their production and sales.

To summarise, contrary to the aims of producers in Lleida when they established modern-day agricultural cooperatives, selling products collectively has not allowed them to negotiate conditions or prices. Joint sales only appear to serve large distributors, allowing them to handle the volumes of stock that they need. Cooperatives allow distributors to outsource the management of producers. Ultimately, agricultural cooperatives have become a tool for disciplining farmers and controlling land to serve the interests of large distributors.

References

Abson, D. J. (2019). The economic drivers and consequences of agricultural specialization. In G. Lemaire, P. C. De Faccio Carvalho, S. Kronberg, & S. Recous (Eds.), *Agroecosystem diversity: Reconciling contemporary agriculture and environmental quality* (pp. 301–315). Elsevier. https://doi.org/10.1016/B978-0-12-811050-8.00019-4

Akram-Lodhi, A. H., & Kay, C. (2010). Surveying the agrarian question (part 2): Current debates and beyond. *Journal of Peasant Studies, 37*(2), 255–284. https://doi.org/10.1080/03066151003594906

Aschemann-Witzel, J., & Zielke, S. (2017). Can't buy me green? A review of consumer perceptions of and behavior toward the price of organic food. *Journal of Consumer Affairs, 51*(1), 211–251. https://doi.org/10.1111/joca.12092

Barciela, C. (2017). La agricultura española desde la Guerra Civil a nuestros días. In G. Carrillo & J. Cuño (Eds.), *Historia agraria y políticas agrarias en España y América Latina desde el siglo XIX hasta nuestros días* (pp. 319–378). Ministerio de Agricultura, Pesca, Alimentación y Medio Ambiente.

Bretón Solo de Zaldívar, V. (2000). *Tierra, Estado y Capitalismo: La transformación agraria del Occidente catalán, 1940–1990.* Milenio.

Campbell, H. (2009). Breaking new ground in food regime theory: Corporate environmentalism, ecological feedbacks and the 'food from somewhere' regime? *Agriculture and Human Values, 26*(4), 309. https://doi.org/10.1007/s10460-009-9215-8

Campos-Climent, V., & Chaves-Ávila, R. (2012). El papel de las cooperativas en la crisis agraria: Estudio empírico aplicado a la agricultura mediterránea española. *Cuadernos de desarrollo rural, 9*(69), 175–194.

Casal, G. (2019). Fruta con justicia social: Cambiar el modelo para recuperar la vida. *Soberanía alimentaria, biodiversidad y culturas, 34*, 32–35.

Clar, E. (2017). La evolución agraria en América Latina y España durante la segunda globalización: 1980–2010. In G. Carrillo & J. Cuño (Eds.), *Historia agraria y políticas agrarias en España y América Latina desde el siglo XIX hasta nuestros días* (pp. 379–430). Ministerio de Agricultura, Pesca, Alimentación y Medio Ambiente.

COAG. (2024). *IPOD: IPOD Índice de precios en origen y destino de los alimentos.* Retrieved June 20, 2024, from https://coag.chil.me/post/ipod-indice-de-precios-en-origen-y-destino-de-los-alimentos-122677

DARP. (2024). *Estadístiques definitives de conreus: Superfícies, rendiments i produccions provincials dels conreus agrícoles.* Retrieved June 20, 2024, from https://agricultura.gencat.cat/ca/departament/estadistiques/agricultura/estadistiques-definitives-conreus/index.html

Farré, M., & Sala, M. (2014). Estrategias empresariales de la gran distribución de fruta fresca en España: Cambios y tendencias. *Anales de Economía Aplicada, 28*, 1410–1425.

García Manrique, E. (1971). La evolución de los regadíos leridanos hacia los cultivos de frutales. *Papeles de Geografía, 3*, 9–134.

Gascón, J., Larrea-Killinger, C., & Solà, C. (2023). Food Waste and Power Relations in the Agri-Food Chain: The Fruit Sector in Lleida (Catalonia, Spain). *Journal of Agricultural and Environmental Ethics, 36*(2), 13. https://doi.org/10.1007/s10806-023-09908-8

Gascón, J., Solà, C., & Larrea-Killinger, C. (2021). *No es negociable: Desperdicio alimentario y relaciones de poder en la cadena agroalimentaria*. Icaria.

Gascón, J., Solà, C., & Larrea-Killinger, C. (2022). A qualitative approach to food loss: The case of the production of fruit in Lleida (Catalonia, Spain). *Agroecology and Sustainable Food Systems, 46*(5), 736–757. https://doi.org/10.1080/21683565.2022.2061099

González, J. A., Garreta, J., & Llevot, N. (2021). Trabajadores temporeros inmigrantes en el campo de Lleida (España): Perfiles y situaciones sociolaborales. *Ager: Revista de Estudios sobre Despoblación y Desarrollo Rural, 31*, 109–137. https://doi.org/10.4422/ager.2021.02

Gonzálves, V., & López Trigal, L. (1999). Jornaleros extranjeros en el campo español. *Ería, 49*, 213–229.

Gorenstein, S. (1998). Sector agroalimentario: Las relaciones industria/gran distribución. *Desarrollo Económico, 38*(149), 457–476. https://doi.org/10.2307/3467388

Iglesias, I., & Casals, E. (2011). Producción, comercialización y consumo de melocotón en España. *Vida Rural, 323*, 27–34.

Isakson, S. R. (2014). Food and finance: The financial transformation of agro-food supply chains. *Journal of Peasant Studies, 41*(5), 749–775. https://doi.org/10.1080/03066150.2013.874340

Juliá, J. F., Meliá, E., & García, G. (2013). Modelos y casos de éxito del cooperativismo agroalimentario en la Unión Europea. *Mediterráneo económico, 24*, 139–154.

Langreo, A. (2012). La estrategia de la gran distribución y su incidencia en la cadena de producción. *Cuadernos de Estudios Agroalimentarios, 4*, 29–46.

Martín, V. O., Jerez, L. M., & Studer, L. (2021). Crisis por cooptación del cooperativismo agrario en Canarias: Una interpretación desde la geografía agraria. *Investigaciones Geográficas, 75*, 203–226. https://doi.org/10.14198/INGEO.17214

Mata, A. (2018). La glocalización y sus consecuencias: Apuntes sobre los temporeros en la fruticultura leridana. *Barataria, 24*, 209–224.

McMichael, P. (2009). A food regime genealogy. *Journal of Peasant Studies, 36*(1), 139–169. https://doi.org/10.1080/03066150902820354

McMichael, P. (2013). *Food regimes and agrarian questions.* Fernwood.
Meliá, E., & Juliá, J. F. (2008). La intercooperación: Una respuesta a las actuales demandas del cooperativismo agrario. *Estudios de Economía Aplicada, 26*(1), 57–88.
Miarnau Doménech, F. (2006). Costes de reconversión y producción para melocotón y nectarina. *Vida Rural, 223*, 26–30.
Miarnau Domènech, F. (2009). Análisis de los costes de producción en melocotón y nectarina. *Vida Rural, 282*, 29–32.
Millán, J. (2009). *Estimació dels costos de producció de fruita fresca a Catalunya: Any 2008.* DARP - Generalitat de Catalunya.
Montagut, X., & Dogliotti, F. (2006). *Alimentos globalizados.* Icaria.
Montserrat, R., Alegre, S., & Dalmau, R. (2004). *Sistemes de formació en presseguer: Avaluació i resultats econòmics als 10 anys d'implantació.* IRTA. Retrieved May 22, 2022, from http://hdl.handle.net/2072/4601
Monzón, J. L. (2003). El cooperativismo en la historia de la literatura económica. *CIRIEC-España: Revista de Economía Pública. Social y Cooperativa, 44*, 9–32.
Observatori de l'agricultura i l'alimentació ecològiques. (2020). *Recull d'estadístiques del sector ecològic a Catalunya: 2000–2019.* DARP-Generalitat de Catalunya. Retrieved May 20, 2022, from http://www.ccpae.org/docs/estadistiques/2019/00_2019_ccpae_recull-estadistiques.pdf
Ploeg, J. D. v. d. (2008). *The new peasantries: Struggles for autonomy and sustainability in an era of empire and globalization.* Routledge.
Prieto, M., Mouwen, J. M., López Puente, S., & Cerdeño, A. (2008). Concepto de calidad en la industria agroalimentaria. *Interciencia, 33*(4), 258–264.
Reardon, T., Timmer, C. P., Barrett, C. B., & Berdegué, J. (2003). The rise of supermarkets in Africa, Asia, and Latin America. *American Journal of Agricultural Economics, 85*(5), 1140–1146.
Robles, B. (2011). La entrevista en profundidad: Una técnica útil dentro del campo antropofísico. *Cuicuilco, 18*(52), 39–49.
Ruiz, M., Sabaté, P., & Badia, C. (2003). La difusión de innovaciones en la industria frigorífica de conservación de fruta dulce no cítrica. *Revista de economía y empresa, 18*(47), 11–24.
Ruiz, M. C., García, E., & Hernández, M. J. (2013). Cómo responden a la crisis económica actual las sociedades cooperativas agrarias: El caso de las almazaras cooperativas andaluzas. *REVESCO: Revista de Estudios Cooperativos, 113*, 120–149. https://doi.org/10.5209/rev_REVE.2014.v113.43384
Ruiz, M. C., Hernández, M. J., & García-Martí, E. (2006). Estado actual de la investigación sobre sociedades cooperativas agrarias en España. *CIRIEC-España: Revista de Economía Pública, Social y Cooperativa, 56*, 65–86.
Sabaté, P., Sabi, X., & Saladrigues, R. (2000). Cooperativas versus sociedades mercantiles: El sector frutero en Lleida. *CIRIEC-España: Revista de economía pública, social y cooperativa, 34*, 51–70.

Segre.com. (2018, May 13). *Lleida ha sacrificat en els últims dotze anys 8.200 hectàrees de pomeres i pereres.* Retrieved May 18, 2022, from https://www.segre.com/noticies/agricultura/2018/05/13/lleida_sacrificat_els_ultims_dotze_anys_200_hectarees_pomeres_pereres_46162_1114.html

Soronellas, M., & Casal, G. (2014). *Pràctiques i estratègies de la pagesia a Catalunya: Trajectòries i plantejaments de futur.* Entrepobles.

Valentinov, V. (2007). Why are cooperatives important in agriculture? An organisational economics perspective. *Journal of Institutional Economics, 3*, 55–69.

Vives-Xiol, J. (2019). El cooperativisme agrari a Catalunya: Orígens i situació actual. *Quaderns Agraris, 46*, 93–112. https://doi.org/10.2436/20.1503.01.98

The Baskets of the Crisis in Marseille (France)

Valeria Siniscalchi

Vegetable Baskets in the Market's Interstices

April 2020: in Marseille, in a square in front of a school closed due to the COVID-19 pandemic, between 6 and 7 p.m., every Monday and Thursday, the members of the *St Giniez basket* line up. Punctual, as always, they are there to collect their "basket of vegetables" that Robert has prepared for them. He has transported his produce to Marseille from his farm in the Luberon, an area located between Avignon and Aix-en-Provence, about 60 kilometers from the regional capital. People keep their distance, as required by the rapidly adopted health regulations, and most wear a mask. We are in the middle of a lockdown, covered and open-air markets are closed, and only a few businesses are allowed to operate, but the prefecture has authorized the distribution of fruit and vegetables such as this one. These are not free distributions, like those organized by national or

V. Siniscalchi (✉)
EHESS, CeRCLEs (Research Center on Circulations, Links, and Exchanges), Marseille, France
e-mail: valeria.siniscalchi@ehess.fr

S. Narotzky et al. (eds.), *Agricultural Extractivism in the Mediterranean Region*,
https://doi.org/10.1007/978-3-032-05599-6_14

international foundations and organizations (*Secours populaire, Caritas*) or by volunteer associations (*AprèsM* for example),[1] that go around the different neighborhoods of the city to help the most fragile and vulnerable people or families who have lost their jobs during the health crisis.

The distribution of fruit and vegetables that takes place in the square of the St Giniez neighborhood is part of a wider system called AMAP (*Association pour le maintien de l'agricuture paysanne*—Associations for the preservation/support of peasant/small-scale agriculture). Similar to the CSA (Community Supported Agriculture) groups in the USA or the Catalan cooperatives that bring together producers and consumers (Homs & Narotzky, 2019) or even the Italian GAS (*Gruppi di Acquisto Solidale*, see Brunori et al., 2011; Grasseni, 2013), AMAPs provide a direct relationship between consumers and producers. There are numerous "anti-middleman" systems, such as the fruit and vegetable distribution networks that developed and spread in Greece after the 2007 economic crisis (Rakopoulos, 2019). Projects and experiments revolving around food become opportunities to create new economic spaces and new ways of thinking about production, exchange, consumption, and the relationships between the individuals connected to these activities (Carrier & Luetchford, 2012; Counihan & Siniscalchi, 2014; Pratt & Luetchford, 2014). In southern Europe in particular, the number of experiments in new ways of producing and/or consuming has multiplied (Siniscalchi & Harper, 2019). Other producers and consumers are building networks that oppose large-scale distribution or attempt to institutionalize alternative experiences—such as the Italian food network that emerged in Italy, *Genuino clandestino* (Koensler & Meloni, 2019)—by promoting their clandestinity in relation to state or European circuits and standards. The works of geographers, sociologists, and anthropologists show the links that exist between situations of crisis and uncertainty and new ways of thinking about or practicing food (Dubuisson-Quellier, 2009; Narotzky, 2012; Narotzky & Besnier, 2014): some appeared precisely during the health crisis due to the coronavirus, others seem to have appeared following the economic crisis of 2007.

[1] *AprèsM* (AfterM) is the name of an ex-McDonald's that closed in 2019 in one of the outskirts of Marseille. It was taken over as a company by a collective of residents and some former fast-food workers to turn it into a social and solidarity-based restaurant. *AprèsM* distributed food boxes during the pandemic.

The AMAPs were born in 2001 in south-eastern France's Provence-Alpes-Côte-d'Azur region, without a direct link to a period of crisis but after a series of scandals that had affected the food production industry and created a desire for greater transparency among certain sections of the population. I am thinking, in particular, of bovine spongiform encephalopathy (BSE)—known as 'mad cow disease'—which started in 1986 and led to cases of transmission to humans in 1996, and of several cases of dioxin contaminations in 1999 that were followed by avian influenza in 2003.[2] At the same time, small producers and family farms producing on a non-industrial scale and/or in organic agriculture had difficulties in finding stable market opportunities due to competition from industrial agriculture and international production, all in a context of European agricultural policies that mainly supported large-scale farming.[3]

Today in France there are a variety of these systems for supplying vegetables and fresh food products. Most of the AMAPs are now part of a national network, but there are also groups that are not affiliated with any local or national association. The common goal of these systems is to support small-scale farming by ensuring that local producers have a market without intermediaries and a guaranteed and stable income (Lamine, 2008a, pp. 22–26). At their heart, they are a critique of the market, the capitalist system itself, and its destructive effects on the economy, values, and the quality of farmers' work and life.

In these pages, I will describe the basket system and the values it promotes, before analyzing the COVID-19 pandemic as a moment of crisis that could have contributed to spreading and consolidating the vegetable basket system, as well as to the forms of adjustment to the initial increase and subsequent decrease in demand. My research on the *Paniers Marseillais* began 15 years ago as part of a broader study on forms of *food activism*. This long-term research allowed me to study the basket system, paying attention to transformations over time, to the internal frictions and divisions. The analyses I present in these pages are the result of observations, participation in meetings and basket distributions, and numerous

[2] In more recent years, the scandals have continued, with, for example, cases in 2013 of horse meat being sold as beef, affecting major companies such as Picard and Findus.

[3] See the classical analysis of Mendras (1967). Rastoin, an agronomist and economist, pointed out as early as 2000 the effects that the growth of the agri-food industry had had on agriculture, particularly what he called the "industrialising technical and economic revolution in agriculture" (2000, p. 65). In the early 1990s, the industrialisation of agriculture and food production seems complete (idem, p. 68).

interviews with producers, with members of the association's board of directors, with volunteers in local groups, and with members in three groups in particular.

The Beginning

Robert recalls the early years of his activity as an organic vegetable producer: "At the beginning I started selling to organic wholesalers, it worked, and there weren't many of us … From 2000 onwards we started to say to each other 'no more wholesalers.' The first year went well, the second as well, and then they started to take fewer products from us and to buy from producers who produce at a lower price." After one of the commercial chains to which he sold salads and broccoli announced that it would only buy salads, Robert saw a truck arriving from Spain unloading crates of broccoli. He said to himself, "I have to find another solution" and then made contact with a group of consumers that had formed in Marseille. An experiment with the "vegetable basket system" was just starting in the city:[4] it was actually "invented" and promoted by producers who conceived it as a way to circumvent wholesalers, traders, and other intermediaries by creating new marketing opportunities. I define the creation of AMAP as an "invention" to emphasize the aspects of design and experimentation that have characterized this system from the very beginning and are still evident today. Participants, in general assemblies or through working groups, regularly reflect on how the system works and how it can be improved. They discuss it as if it were something created locally and developed over time through adjustments and attempts to "improve" it, although similar systems already existed elsewhere.

Over the years the groups have multiplied and given rise to a network, *Alliance Provence*, from which the *Paniers Marseillais* (Marseille baskets) association emerged in the mid-2000s, as the result of an internal division. When it was created, this association chose to emphasize not only local, small-scale farming, like the other AMAPs, but also production respecting organic farming regulations (which were not required in the AMAP

[4] Even if their inspiration came from the concept of Community Supported Agriculture, which some of the producers in the area of Marseille had discovered in the US (Lamine, 2008a, pp. 17–21), I use the term "vegetable basket system," similar to the French term "*système de paniers de legumes,*" in order to underline the specificities of this provisioning system as compared to other similar systems.

system). Another characteristic of the *Paniers Marseillais* is that the network only covers the city of Marseille (groups in other municipalities in the region continue to be part of *Alliance Provence*, now *Amap de Provence*). The function of the *Paniers Marseillais* association is to support and facilitate the creation of local groups and to maintain the link with fruit and vegetable producers and other "local" producers for which the association coordinates the orders and distribution of products (meat, cheese, flour, citrus fruits, etc.) that members can choose to purchase in addition to the vegetable basket (once a month or every three months).

A Vocabulary Against Agroindustry

The *Paniers Marseillais* association (the acronym, PAMA, is also an anagram of AMAP) consists of around 30 neighborhood groups, each made up of a variable number of members (between 20 and 80) corresponding to the same number of households. Each group is linked to a market gardener/farmer.[5] An annual contract, referred to as a "solidarity contract," is established between each member and the farmer: it establishes the member's advance purchase of the farmer's production, and the weekly distribution of baskets of vegetables by the farmer. When "solidarity" is introduced in exchange relationships, it identifies a set of practices and attitudes conceived as tools to oppose and escape the competitiveness and power relations that characterize the market system and industrial agriculture. Solidarity in "alternative" systems has an impact on the conception of the terms of exchange, the relationship to money, and the form of the contract. It embodies an ideal of balance and allows for the enhancement of the social dimension of exchange, conceived as an element to be preserved or reintroduced when it seems to disappear or progressively diminish in food provisioning (Siniscalchi, 2019). In the food distribution groups that were set up during the economic crisis in Greece, for example, solidarity takes the form of a network of activities based on reciprocity and a set of obligations that aim to replace market exchange relationships. It constitutes a concrete response in difficult situations and at the same time a counter-hegemonic discourse to the language of debt, particularly

[5] One of the first published works on the AMAP system is the book by sociologist Claire Lamine (2008b) based on an ethnography in south-eastern France. See also Poulot (2014). My description of the AMAP system here is based on parts of articles I published previously (Siniscalchi, 2019, 2021, 2025; Consalès et al. 2022).

dominant at the time, a way of building a more democratic economy (Rakopoulos, 2016). In the case of vegetable baskets, the "solidarity" dimension is considered the central and constitutive element of the contract established between farmer and consumers. Solidarity refers to the sharing of risks due to the unpredictability of production: in the event of a poor harvest or destructive weather, the consumers accept that their share of vegetables could decrease. The risk is shared by paying for the baskets in advance and signing a one-year contract. The price of the basket indicated in the contract is fixed and independent of variations in the quantity and weight of the vegetables, which is not specified. It corresponds to the average weekly "value" of the basket, that means the economic value, but also its environmental and political value. In this protected social space, "economic proximity" means the possibility to discuss economic questions: the price is set in agreement with the members of the group, but at the same time, each producer harmonizes the price of his baskets with that of the other producers. This price is generally cheaper than the market price paid by consumers for similar products in supermarkets or shops but higher than the typical wholesale price received by producers. Although quantity and weight are not fixed, many contracts do indicate the number of varieties included in order to avoid the risk of members receiving a basket containing, for example, only cauliflower or only potatoes at certain times of the year. Once the production year has been planned, the producer must forecast and calculate the approximate quantity and variety of vegetables which will fill the baskets throughout the 52 weeks (or 48 if the contract includes a few weeks' break).

Farmers can distribute their products to one or more local groups, depending on the production capacity of their farm and the number of members in each group. Each group, however, is linked to a single vegetable producer. In the "ideal" model, the annual production of one farmer is exclusively destined for the basket members, even if in reality, half of the farmers who are members of the *Paniers Marseillais* sell part of their production directly on the farm or through farmers' markets, effectively combining multiple systems. The farmers who join the basket system come from different backgrounds and professional experiences: some own the land and acquired the necessary skills from their farming families, even if they often worked in other jobs before returning to the family land. Others have settled more recently and acquired the land or part of it through loans. Producers who have contracts in the vegetable basket system must adjust their production quantities and logic to the system, moving from monoculture to varied market gardening, and adding distribution to their production activities. As economic situations vary, so do production choices and market channels. In any case, for the system to work and maintain a balance for farmers and consumers, those farmers who decide to maintain or develop other markets in parallel with AMAPs must be economically stable and willing to integrate the AMAP "spirit" so as not to disadvantage members in favor of more volatile market opportunities.

The logic behind this economic relationship, imagined as "exclusive," is linked to the variation in production at different times of the year. This means that the baskets contain different quantities (and varieties) of products from one week to the next, and the balance between the price paid for each basket and its contents is achieved over the whole year and not for every single week.[6] When the vegetable basket system is the producer's only source of income, he or she will plan the annual production based on the number of members in his or her group or groups, and of course, the number of members must not exceed the farm's capacity: periods when the contents of the baskets are more abundant (such as summer and some winter months) will compensate for the months between two seasons when the baskets may be less full.[7] Members look for quality products, but they are also seeking more transparency in production, traceability, and less transportation. They are interested in reducing intermediaries and having a direct relationship with the producer. Direct exchange, solidarity, and proximity (geographical and social) are the main issues they defend, and the exclusive bond between a group and a producer contributes to create and maintain this social proximity. Some producers have reciprocal relationships with other producers in the network from whom they occasionally purchase (or exchange) certain products to complete their baskets (e.g., when they experience a drop in production due to weather conditions or when they have particular difficulties with a type of vegetable).

The farmer often defines himself or herself as a "peasant": the term peasant, which had long been devalued and associated with the negative images of "backward" land workers, is now used in this context to emphasize that AMAP producers are not focused on the same capitalistic goals

[6] The basket contains whatever seasonal products correspond to the calendar of crops established by the farmer in the contract: approximately six or seven different kinds of products according to some contracts.

[7] Farmers who produce exclusively for the groups of members, without other outlets, went through a "mixed" phase at the beginning of their participation in the system, with a gradual transition to the "ideal" model.

found in industrial agriculture. The revaluation of the term peasant is taking place in many other contexts, see La Via Campesina for example. "Consomm'acteur" is the term used to refer to consumers in official AMAP documents (contracts, charters). This neologism indicates that consumers have an active role: not only through their mobilization during the distribution of baskets but also as actors aware of their consumption choices. "We haven't found a nice word: adherent isn't nice, consumer, even worse! So, friends, I call them friends, we haven't found any extraordinary words," Robert reminds us.

The term *"panier"*—a derivative of the French word for basket—indicates the quantity of vegetables to which each member of the association is entitled on a weekly basis (usually couples or families with children, but sometimes single people). Each local group, with between 20 and 70 members, is also called a *panier* (basket) and is named after the neighborhood or place where the weekly distribution takes place, resulting in names such as, *Panier d'Endoume, Panier de Saint Giniez, Panier du Rouet.* These places sometimes resemble a small market square due to the way the tables are set up with vegetable boxes and scales where the group's volunteers weigh some of the products. The members form a line at the tables when it's time to collect the vegetables, one by one, to fill the basket. This set of activities is referred to as "distribution."

Being a subscription based system, the monetary transaction is invisible at the place of distribution. Members simply sign next to their name on a list for which some volunteers are responsible. Payments have already been made by checks given in advance to the association's treasurer, who is in charge of collecting contracts and payments and delivers the checks to the producer every month. In recent years, most groups have replaced checks by an automatic bank transfer scheduled to be sent on a fixed date each month. The invisibility of money during the basket distribution allows members to think of the system as something other than a market, and not just through the use of vocabulary. Compared to a market or supermarket, the system also involves obligations that make it more rigid than other supply systems: the members cannot choose what to put in their basket (the baskets are already put together by the farmer or are put together at the time of distribution according to the list and quantities indicated by the producer), nor can they choose the best-looking products, and they must come and collect their basket every week at the appointed time. When a member cannot collect his or her basket or goes on holiday, he or she can exchange it or resell it within or outside the group, or can send someone to collect it for them. If a member has to terminate the contract, for example, due to a relocation, they undertake to find a replacement who will take over their contract. These clauses are an integral part of the solidarity contract and are intended to ensure continuity and economic stability for producers, unlike what happens, for example, in the competitive system of large-scale distribution.

Pandemic

New practices emerged during the food basket distributions that took place throughout the lockdown. Since the courtyard of the social center where the distribution usually took place was closed due to health restrictions, the group members of *St Giniez basket* had found a space that was open, but not very visible from the street. AMAP distributions were authorized by the prefecture (in accordance with the decrees of March 16 and 25, 2020 that gave prefects the freedom to assess different situations) but everyone was afraid that the rules might change, so they thought it would be better to avoid attracting the attention of the neighborhood. When the group members came to collect their baskets, the first thing they did was to put a check or a sealed envelope containing the payment for their basket inside a larger envelope that the association's treasurer handed to them. Robert, the producer, did not receive the money directly. A volunteer member of the association was in charge of collecting the payments, while other volunteers helped put together the baskets. The money was neither counted nor handled on site. But these remote operations and the lack of visibility of the money were not due to the health situation. On the contrary, during this period the money circulated at the distribution points more than on the evenings of any other month of the year: although hidden in the envelopes, it was more in evidence than usual. Why was this? Members' baskets are normally paid for in advance, and this should have been a month off for Robert, one in which no distributions were planned. But the high demand for fresh fruit and vegetables due to the health emergency and the closure of open-air markets convinced Robert and the other farmers who are members of the *Paniers Marseillais* network to forego their time off and ensure distributions. This unplanned month of distributions meant that members would need to pay for their basket on the day of distribution.

Thierry also normally distributes his baskets in the courtyard of a social center, on Thursday evenings. But like Robert's group, the members of

Thierry's group, *Endoume basket*, had to find a place to replace the closed social center. They found a small cul-de-sac away from any busy roads, even during the lockdown, in order to be less visible. Thierry was among the farmers capable of distributing more vegetables, and members spread the word, which led some of their neighbors or friends to reserve a basket. The system offers the possibility of selling additional baskets during the distribution as a way for interested people to try the system and for the association to reach new potential members. Before the pandemic, the demand for these "test" baskets from non-members was not frequent. During the health crisis, however, according to several producers and the media, the demand seemed to have increased, and not only in the city of Marseille. In fact, almost everywhere in France, the period of the pandemic coincided with a set of individual and collective restrictions that forced people to change their daily rhythms and habits, particularly those related to food. The period of lockdown restricted the number of markets people could access, and the operations of some open-air fruit and vegetable markets were suspended. At the same time, the closure of bars and restaurants, the need to work from home, the restrictions on the time spent outside the home, and the temporary closure of schools and school canteens, all encouraged or obliged many of those who usually ate at least one meal outside the home to start cooking at home. The sharing of recipes and enthusiasm for the fresh vegetables, combined with the fear of possible food shortages, attracted new consumers to the vegetable basket system. Alongside the "test" baskets, some groups received membership requests from new consumers. Newspapers and online blogs started to report this growing interest in local food, new consumers seem to have discovered the existence of the system and the qualities of organic food.[8] This fits in with a general impression, conveyed by the media, of a renewed appreciation of values other than consumerism, of a reduction in waste and excess: "Between market closures, health risks and shortages of certain

[8] See among others: "Pendant le confinement, les AMAP s'organisent pour continuer les distributions de fruits et légumes" Le Monde, March 25, 2020 (https://www.lemonde.fr/m-perso/article/2020/03/25/coronavirus-les-amap-s-organisent-pour-continuer-les-distributions-de-fruits-et-legumes_6034428_4497916.html); "Coronavirus: les AMAP attirent de nouveaux consommateurs avec leurs abonnements à des paniers frais" Francebleu April 18, 2020 (https://www.francebleu.fr/infos/agriculture-peche/coronavirus-regain-d-interet-pour-les-amap-et-les-producteurs-locaux-1586339068); "Les Amap, îlots de lien social dans l'océan du confinement" Reporterre, March 27, 2020 (https://reporterre.net/Les-Amap-ilots-de-lien-social-dans-l-ocean-du-confinement).

foods, the assurance of receiving a weekly basket of fresh produce is appealing to some consumers and giving others something to think about, according to ... [the] president of the AMAP network in Savoie." (https://www.francebleu.fr/infos/agriculture-peche/coronavirus-regain-d-interet-pour-les-amap-et-les-producteurs-locaux-1586339068). "Unprecedented! says Chantal, who hopes that this global health crisis will raise awareness among consumers and those in positions of power. Prioritizing short supply chains and supporting and maintaining local producers while minimizing risks is part of the solution." (https://www.ouest-france.fr/sante/virus/coronavirus/coronavirus-fermer-les-amap-serait-contre-productif-6800769).

For a few weeks or months, the number of members did seem to increase. But in reality, the demand that the producers of the *Panier Marseillais* received was on average five, maximum ten baskets more. For those producers who lost other commercial outlets (such as markets), this surplus allowed them to reduce their losses. For those who dedicate themselves exclusively to basket production and were already in a position to take on new members because they had not yet reached the maximum number of baskets/members, this surplus was welcome: it gave them hope that the number of members might increase even after the end of the pandemic. Once the health crisis was over, however, urban consumers' interest in direct forms of provisioning seems to have fallen back to pre-pandemic levels, forcing producers into new forms of adjustment. Why did this interest diminish when the pandemic period passed?

And After the Pandemic...

The vegetable baskets were a sort of adjustment factor for the consumers "passing through" the basket system. An adjustment variable not so much in economic terms (although the price of vegetables is generally lower than the cost of the same quantities and varieties of products in a supermarket or greengrocer's due to the absence of intermediaries), but rather in the social practices of consumers: more time spent at home and dedicated to food preparation. These new basket consumers, who discovered the system via networks of acquaintances and friends, benefitted from the "attractive" side of the AMAP system (fresh, quality, and tasty produce delivered not far from home), but their approach to the system was only temporary, as was the rediscovery of home cooking for many consumers.

This is probably due to the structural characteristics of the basket system. Subscribing to a vegetable basket, in normal times, requires a change in habits, particularly culinary habits. The basket must be collected once a week, the vegetables washed and prepared: they are not individually selected by the consumer, they are often earthy and not ready to be cooked or consumed like vegetables bought in shops or supermarkets. Furthermore, for many members, a subscription to a vegetable basket increases the quantity of vegetables consumed compared to what they would buy through "traditional" means of provisioning. For occasional consumers, buying vegetable baskets during the pandemic didn't require a radical change in habits, which had already been modified by the pandemic: long periods at home, the restrictions to shop only within a limited distance, the need for larger quantities of food because the children didn't eat at school, and the closure of places where people usually eat prepared food (canteens, restaurants). These conditions constituted an indirect incentive to cook at home.

Most of those who experimented with the baskets "thanks to" or "because of" the pandemic, were able to do so because they had a surplus of time and there was a surplus in the available offer from some producers. In turn, these producers, or at least some of them who had lost their normal complementary sales outside of the basket system, were thus able to compensate for the loss of income and maintain a balance, even if a precarious one, thanks to the increased sales of vegetable baskets. But what happened after the pandemic? Why didn't the "test" baskets bring in new consumers on a more stable basis? The gradual reopening of traditional points of sale and the return to normal working patterns brought the system back to the pre-pandemic situation. Of course, the AMAP system has shown itself to be "resilient," as Gassama and Lagane (2022) claim, but this elasticity and ability to adapt on the part of producers and members would not have been possible without a favorable regulatory environment (distributions were authorized by many prefectures). Furthermore, the flexibility during the pandemic, which was partly real and partly ostensible, has given way to the inherent rigidity of a system which might not appeal to "passing consumers." The main constraints of the system—according to those who don't renew their subscription—are in fact the need to collect the basket almost every week of the year or to find another interested person to collect the basket during one's absences: "vegetables grow even during the holidays" as many producers tell the basket members. For the *consumm'actors* who adhere to these same systems, it means (re)learning to cook, familiarizing themselves with varieties of vegetables that are not available in supermarkets, with tastes that are very different, and having to deal with the constraints of the seasons and irregular supplies in quantity. The necessity to dedicate time to preparing food plus the impossibility of choosing products in terms of variety and quantity and finally accepting that at certain times the baskets are less abundant all contribute to making the system too complicated for some people. If these elements dissuade new consumers from joining the system or drive away some members who decide not to renew their subscription (each local group sees a turnover from one year to the next), the duration of the contract at the same time allows others, who have joined for different reasons (for fashion, to do as their friends do, for the apparent practicality, for health reasons) and signed one year contract to slowly modify their own practices and habits and to develop an ideological adherence to the system.

Conclusion

Changing practices also means changing paradigms and vice versa (Siniscalchi, 2015). As I have emphasized elsewhere, the capacity for change in these experiences is triggered at the crossing between the scale of individual action and the collective dimension that characterizes them (Siniscalchi, 2025): those who seem to have simply "passed through" the system (when the passage was longer than just a few weeks) and then returned to more traditional practices do not come out unscathed. Proximity and curiosity are at the origin of the adhesion of certain families to the vegetable baskets system. But a simple weekly vegetable basket for a few months also familiarizes people with other issues at the heart of the system, such as the viability of small farms, the quality and traceability of food, environmental protection, the health of producers, and the accessibility of the system. The attention paid to food, including that from a sensory and taste perspective, is often no longer the same. Similar processes appear within the Catalan cooperatives studied by Patricia Homs and Susana Narotzky (2019, p. 137): "One of the most relevant changes among consumers participating in food coops is the modification of their motivation to take part in them: Most people get involved in a food coop in search of cheaper and healthier food; yet, after some time as members of the group, they redefine their motivations and interests in political terms, valuing aspects such as the kind of relations of production existing among farmers, or the social significance of collective self-provisioning of food." And again "these groups do not present an 'alternative' *model* for social change. Instead, they present a performative *alternative*: practices change people and transform their values through daily activities such as eating, buying, decision-making, and so on." (ibid.)

Furthermore, the basket distribution usually takes place in a familiar environment, where the members discuss among themselves and with the producer or with the volunteers who collect the signatures of those who have collected their basket. The distribution is a social event creating both a "social" and an "economic proximity" (Siniscalchi, 2019). During the pandemic, however, the rules of distancing, wearing masks, and above all, the limited time to devote to the distribution coupled with an injunction against forming groups of conversation had suppressed the sociability that normally characterizes them. Under normal circumstances, that sociability constitutes another attractive element that confirms to members that other methods of supply and exchange are possible. Looking back at the attention being paid to different types of food and provisioning during the pandemic, we see that people broke with their usual habits and perhaps questioned them during a "suspended" but brief time period. But once activities and "normalcy" resumed after the end of the health crisis, the material conditions returned to what they were before the crisis.

Acknowledgments The research was funded by the Direction de l'Économie Régionale, de l'Innovation et de l'Enseignement Supérieur, Région PACA through two successive projects (*DEVAMAP. Quel développement pour les AMAP?*, 2009–2011 and *EQUALIM-TERR. Circuits courts alimentaires en Région PACA. Qualité, équité et gouvernance territoriale*, 2012–2014). Nicoletta Stendardo participated in some phases of the project in 2011 and 2012 as a contract research assistant.

References

Brunori, G., Rossi, A., & Guidi, F. (2011). On the new social relations around and beyond food. analysing consumers' role and action in Gruppi di Acquisto Solidale (Solidarity purchasing groups). *Sociologia ruralis, 52, 1*, 1–30.

Carrier, J. G., & Luetchford, P. (Eds.). (2012). *Ethical consumption: Social value and economic practice*. Berghahn.

Consalès, J. N., Guiraud, N., & Siniscalchi, V. (2022). Les expériences du « local ». Variations et tensions autour de l'alimentation locale dans la région marseillaise. *Nature, Science, Société (NSS), 30*(1), 58–71.

Counihan, C., & Siniscalchi, V. (Eds.). (2014). *Food activism: Agency, democracy and economy*. Bloomsbury.

Dubuisson-Quellier, S. (2009). *La consommation engagée*. Sciences Po, Les Presses.

Gassama, E. B., & Lagane, J. (2022). Vers une société qui se réinvente...Quelles réponses des AMAP en temps de pandémie ?. In N. Chapon & L. Mucchielli (Eds.), Le confinement. Genèse, impacts et controverses (pp.167-188). Presses universitaires de Provence.

Grasseni, C. (2013). *Beyond alternative food networks. Italy's solidarity purchase groups*. Bloomsbury.

Homs, P., & Narotzky, S. (2019). Within and beyond the market system. Organic Food cooperatives in Catalonia. In V. Siniscalchi & K. Harper (Eds.), *Food values in Europe* (pp. 132–146) Bloomsbury.

Koensler, A., & Meloni, P. (2019). *Antropologia dell'alimentazione. Produzione, consumo, movimenti sociali*. Carocci.

Lamine, C. (2008a). *Les AMAP: Un nouveau pacte entre producteurs et consommateurs?* Yves Michel.

Lamine, C. (2008b). *Les intermittents du bio. Pour une sociologie pragmatique des choix alimentaires émergents*. Éditions de la Maison des sciences de l'homme.

Mendras, H. (1967). *La Fin des paysans*. SÉDÉIS, 1967 (réédition Actes-Sud, 1992).

Narotzky, S. (2012). Alternatives to expanded accumulation and the anthropological imagination turning necessity into a challenge to capitalism? In B. P. Gardiner, B. Leach, & W. Lem (Eds.), *Confronting capital. Critique and engagement in anthropology*. Routledge.

Narotzky, S., & Besnier, N. (2014). Crisis, value, hope: rethinking the economy. *Current Anthropology, 55*(S9), 4–16.

Poulot, M. (2014). Histoires d'AMAP franciliennes. Quand manger met le local en tous ses états. *Territoire en Mouvement, 22*, 40–53.

Pratt, J., & Luetchford, P. (2014). *Food for change: The politics and values of social movements*. Pluto Press.

Rakopoulos, T. (2016). Solidarity: The Egalitarian tensions of a bridge-concept. *Social Anthropology, 24*(2), 142–151.

Rakopoulos, T. (2019). 70% Zapatista? Solidarity "ecosystems" and the troubles of valuing labor in food cooperatives. In V. Siniscalchi & K. Harper (Eds.), *Food values in Europe* (pp. 147–160). Bloomsbury.

Rastoin, J. L. (2000). Une brève histoire de l'industrie alimentaire. *Économie rurale., 255–256*, 61–71.

Siniscalchi, V. (2015). 'Food Activism' en Europe: changer de pratiques, changer de paradigms. *Anthropology of food* [Online], S11, http://journals.openedition.org/aof/7920

Siniscalchi, V. (2019). Solidarity, calculation, and the economic proximity inside the "vegetable basket" system in the South of France. In V. Siniscalchi & K. Harper (Eds.), *Food values in Europe*. Bloomsbury.
Siniscalchi, V. (2021). Il denaro invisibile. Circolazione, calcolo e valori negli spazi economici dell'attivismo. *Cheiron, 1–2* (2019), 252–275, FrancoAngeli.
Siniscalchi, V. (2025). Mobilisations et 'alternatives' économiques: la nourriture comme objet politique. In B. Hubert (Ed.), *Actes des colloques de Cerisy* (Forthcoming).
Siniscalchi, V., & Harper, K. (Eds.). (2019). *Food values in Europe*. Bloomsbury.

Index[1]

[1] Note: Page numbers followed by 'n' refer to notes.

S. Narotzky et al. (eds.), *Agricultural Extractivism in the Mediterranean Region*,
https://doi.org/10.1007/978-3-032-05599-6

C

D

N

O

P

Q

R

Kolibri 360 GmbH
Ferdinand-Jühlke-Straße 7
99095 Erfurt, Deutschland
produktsicherheit@kolibri360.de